Science at the Cutting Edge

The Future of

the ØRESUND REGION

Science at the Cutting Edge

The Future of

the ØRESUND REGION

Gunnar Törnqvist

Science at the Cutting Edge. The Future of the Øresund Region

© *Copenhagen Business School Press*
Book designed by Elisabeth M-Pyrko, Formgivning Art2Di2
Cover designed by Elisabeth M-Pyrko
Printed in Denmark by Narayana Press, Gylling
1. edition 2002

ISBN 87-630-0101-2

Distribution:

Scandinavia
Djoef/DBK, Siljangade 2-8, P.O. Box 1731
DK-2300 Copenhagen S, Denmark
Phone: +45 3269 7788, fax: +45 3269 7789

North America
Copenhagen Business School Press
Books International Inc.
P.O. Box 605
Herndon, VA 20172-0605, USA
Phone: +1 703 661 1500, fax: +1 703 661 1501

Rest of the World
Marston Book Services, P.O. Box 269
Abingdon, Oxfordshire, OX14 4YN, UK
Phone: +44 (0) 1235 465500, fax: +44 (0) 1235 465555
E-mail Direct Customers: direct.order@marston.co.uk
E-mail Booksellers: trade.order@marston.co.uk

FOREWORD

To promote Danish-Swedish research co-operation within and for the Øresund region the governments of Denmark and Sweden have decided to establish a jointly financed Committee for Research and Development of the Øresund region, *Øforsk*. The Committee supports joint research initiatives at the most advanced scientific level. Moreover the Committee may reflect on consequences of such research for society. A recent question of that type concerns regional consequences of the location of major research establishments, e.g. the proposed European Spallation Source (ESS), meant to be built somewhere in Europe. What would be the effects upon the Øresund region if such a facility was located to the region?

Øforsk is proud that prof. dr. Gunnar Törnqvist accepted to perform this analysis, which may be of great general interest, not just for the Øresund region. Thereby the analysis also may be of value as a textbook for higher education.

Lars Pallesen
Chairman of *Øforsk*

CONTENTS

1

Assignment

Several of the conditions governing this report have been laid down in advance. Its primary task has been to discuss the probable local and regional impacts of the possible establishment of a major research centre. The impact of this research centre on the surrounding society forms the focus of the report. Within this framework, the author has then used his own special competence to define the issues to be examined.

The report is primarily concerned with the Øresund region. This provides an opportunity for the author to refer to previous research that he has carried out on this region. However it is the author's intention that neither the choice of geographical area nor the type of institution involved should stand in the way of a general discussion of the probable effects of the establishment of a major research centre. Considerable attention will be paid to comparable international surveys. This presentation will tend to reflect the view that in the present era, a region's particular characteristics and future potential cannot be assessed without taking account of the forces operating in a changing world.

A relatively short period of time has been available for the completion of this project. Consequently it has not been possible to carry out a major empirical investigation. Accordingly we have largely concentrated on summarising previous studies and analysing the fairly comprehensive literature that is available in this field. Finally the author has been able to draw on his recent experience of working on a major international research programme that is in the process of completion.[1]

In its present form, this book is primarily intended to provide a basis for discussion and future research. Although they are soundly based, several of the ideas presented in the report will require to be further developed once concrete investment plans have been formulated.

Background

A fixed link between Copenhagen and Malmö has been in operation since July 2000. It is a combination of bridge and tunnel that caters for both road and rail transport. It took three years to build and is 16 kilometres long. Charges will in the long run cover the costs of the investment loans. The approach roads that have been and are in the process of

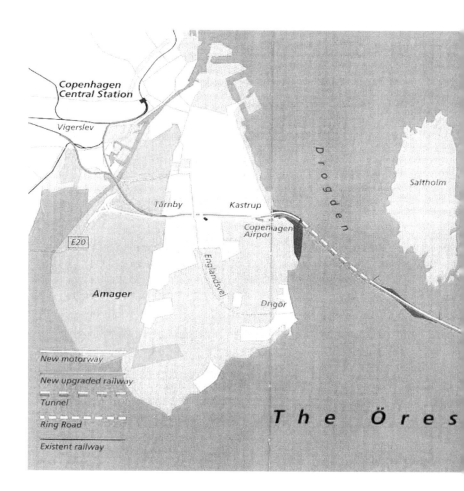

Figure 1 The fixed link between Copenhagen and Malmö

construction have helped to increase *physical accessibility* within a geographical area comprising three million people and approximately half as many places of work (Figures 1 and 2). This is a considerable agglomeration by Scandinavian standards. This improvement in communications should be seen as an important step in the direction of

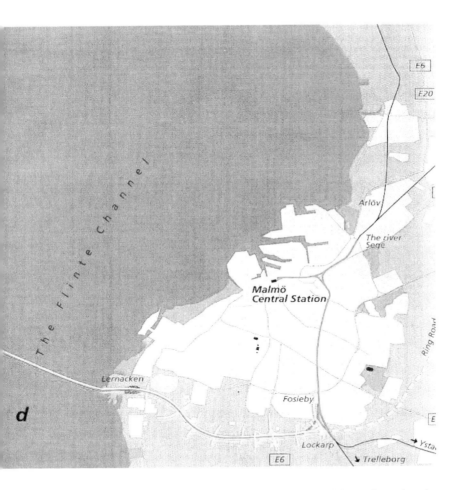

creating a strong regional unit on both sides of an old state boundary in Northern Europe.

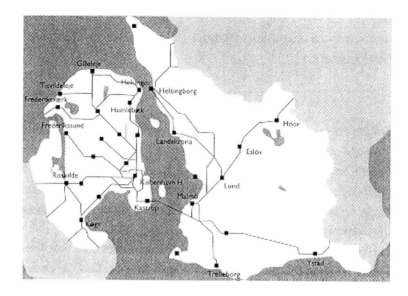

Figure 2 **Direct rail communications in the Øresund area at the beginning of 2000**

The bridge

The question of a fixed link between Denmark and Sweden has been discussed and analysed for more than one hundred years. The technical solutions put forward have ranged from tunnels and various bridge projects to a proposal to drain Øresund, thereby reclaiming valuable land at the same time as a fixed link was established.

The planned link has in various periods acted as a *symbol* for the spirit of the time and its prevailing political views. It has been seen as an expression for the technical progress of the period and the current state of the art of civil engineering. From an early stage, it was seen as a symbol of Nordic brotherhood and subsequently as an image of the successful Swedish export industry's growing need for improved transport facilities. In the 1960s, the supporters of a bridge across Øresund developed the vision of an Ørestad, a technopolis and a European industrial metropolis. The opponents of the bridge project viewed it in terms of a new Ruhr area.

In the 1970s, the bridge was increasingly seen together with the nuclear power station at Barsebäck as a serious threat to the environment, not least on the Danish side. Prior to the construction of the bridge and tunnel, a great deal of attention has been devoted to the environmental effects of the project, especially its effect on the circulation of water through Øresund and on animal life. However the construction of this fixed link has been undertaken with great care and has undoubtedly reduced the possibility of environmental damage. The environmental risks associated with an expansion of traffic in the Øresund region is to some extent a different issue which will undoubtedly be discussed in the years ahead.

The evident symbolic value of the fixed link has been apparent both during and after its construction. It is obvious that the bridge especially provides a concrete and highly visible structure that attracts interest and in a rare sense acts as a source of inspiration. Discussions are now taking place that could naturally have occurred in the absence of the fixed link. It is hardly credible that the standard of communications across Øresund prior to the fixed link was so deficient as to have become the single decisive factor determining the low degree of integration in the region.

At the same time as the fixed link raises questions regarding the effects of major infrastructure investment programmes, it also focuses on issues concerning the opportunity to create viable integrated regions separated by a boundary between old sovereign states. It is a widely held view that the Øresund area has excellent opportunities to participate in the competition between the dynamic regions of Europe.

The bridge and the nascent regional integration may be seen as a gigantic and internationally unique social experiment in the most densely populated region in Scandinavia. This experiment is likely to have far-reaching consequences for the natural environment, human settlement, the economy, legal system, administration and culture on the periphery of two relatively closed spheres of national interest.

Since the opening of the bridge, an air of impatience and frustration has tended to predominate in the public debate. The volume of traffic across Øresund has not developed as planned. While the number of rail passengers has exceed expectations, the growth of road transport has been slower than forecast. From an environmental standpoint, this is naturally beneficial. However the advocates of a strong, integrated Øresund region have in this situation become increasingly concerned with the numerous *non*-physical barriers that are associated with the boundary between Denmark and Sweden. Let us pause to examine what we know about these barriers.

The state boundary

Companies on the Swedish and Danish sides of Øresund have had limited trade and remarkably few internal business contacts across the border. The networks of sub-contractors as well as the distribution systems in both Zealand and Scania have been closed to one another. The innovation systems that diffuse technology and new ideas within the respective economies have also remained isolated from each other. The barrier effect is most significant in the actual border areas. Since the Second World War, there has been a marked expansion of the total volume of foreign trade between Sweden and Denmark as well as between Sweden and the Continent. This is borne out by the growth of *transit* traffic in Helsingborg, Malmö and Trelleborg. However the trade and contacts that are aimed at points within the border area are strikingly limited. Although most of the major studies concerning barrier effects were carried out almost twenty years ago, there are numerous indications from more limited studies which support the view that little has changed in the region during the period up to 2000. Significant barriers remain to be penetrated.[2]

Passenger transport across Øresund has been almost entirely a matter of journeys to and from the Copenhagen airport Kastrup, and shopping trips related to price differences for particular goods. Commuting has increased markedly during the 1990s from a figure of several hundred in the 1970s to around 3 000 in the 1990s. By way of comparison, there are almost 300 000 commuters that travel along the Danish side of Øresund.

The studies that have been carried out indicate that the lack of business contacts across Øresund is not just a question of transport difficulties. Nor do fiscal, technical and other barriers to trade appear to present problems in the cross-border region. It is possible that language differences act as a barrier. However it would appear that the state boundary per se and all that it implies together with the post-war shift towards national centres has been the primary obstacle to the growth of a functioning Øresund region. The routines encompassed by the administrative regulatory system, ownership conditions, network structures and organisational forms have shifted development away from the border regions. On the Danish side of Øresund, Copenhagen has for a long time acted as a focal point for decision-making in the Danish economy as well as an important node in the international contact system. The overwhelming proportion of Copenhagen's administrative and commercial interests are to be found to the west of the Danish capital. By comparison with Copenhagen, Malmö and Helsingborg are typical branch regions. Here relationships are directed towards the north-east and the heartland of the Swedish nation state. Firms in Malmö and Helsingborg have for a long time had ten to twenty times more contacts (trade, business trips and telephone contacts) with Stockholm than with their own backyard on the other side of Øresund. Many of the head offices of big Swedish companies are located in Stockholm. From there they handle the international contacts of importance. In an international perspective governed by strong sovereign nation states, only Stockholm and Copenhagen are on the same hierarchical level.

Following the Second World War and the division of Europe, the central location of the Øresund region as a crossing point for foreign trade in Northern Europe was redefined. Moreover it is also the case that Malmö and Copenhagen have tended to view themselves as located in problem areas lacking support from the central authority despite the fact that Copenhagen is a capital on the Danish side. As we shall see below, there are numerous peripheral regions in Europe that experience their predicament in exactly the same way.

Since 1960, Denmark's industrial centre of gravity has gradually moved westwards. This shift has been due to the growth of existing companies and a concentration of new companies to Funen and Jutland rather than companies moving out of the Copenhagen area.[3] The industrial city of Malmö lost its textile and clothing industries and subsequently its major shipyard. A motor vehicle plant proved to be a short-lived solution. In a sparsely populated country full of areas losing population, it was difficult for the Swedish government to see the densely populated south-western Scania as a region in urgent need of aid.

Time

Nowadays, once decisions are finally made, new transport and communication systems are rapidly built. Other changes often take longer to implement. Against the background of long experience of development projects, it has been said that "it is easier to build a road than to create

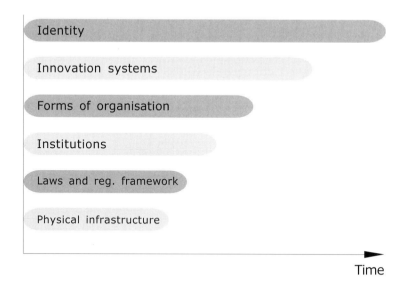

Time

Figure 3 Propensity to change over time. Outline

an efficient organisation to maintain the road". Institutional changes take time. People have to live with a new institution for a long time before its special effects on behaviour and cultural conditions start to make an impact (see the schematic presentation in Figure 3).

Generally speaking, different laws, rules and regulations apply on different sides of a state boundary. It may cover everything from tax legislation, law of contract and labour law at the national level to traffic regulations and opening hours at the regional and local levels. It would seem to take longer to harmonise entire legal frameworks than to build a bridge.

This process is however somewhat simplified by the fact that Sweden and Denmark are both members of the European Union. It will nevertheless take considerable time before these measures have any appreciable impact on human behaviour.

The future development of the Øresund region will also be dependent on the local and regional authorities that enter into legitimate, binding agreements with each other and the degree to which it becomes possible to create common decision-making bodies over time. The combined territorial reach of the decisions and regulatory frameworks for these common decision-making units will be one of the ways in which the boundaries of the region are actually determined in the future.

The creation of common, powerful decision-making bodies is obviously not a simple matter. There are numerous examples of conflicts between local authorities and various regional groupings on both sides of Øresund. Regional and local rivalries, particularly on the Swedish side of the Sound, have tended to impede collaboration between authorities in relation to various types of projects. National decisions on a fixed link across Øresund were held up by local disputes regarding the actual location of the fixed link. There is also the matter of the extent to which the Swedish and Danish central governments are prepared to delegate authority. There is a considerable risk that a cross-border regional authority will find itself in a *constitutional vacuum* in its relationship with central governments. As may be seen throughout Europe in relation to cross-border regions, the attempt to create common institutions has been conducted via agencies. We will examine this process in greater detail below. Various agencies and chambers of commerce seek to co-ordinate activities. Local authorities and regional councils are represented on these bodies. Trades unions, employment agencies and cultural organisations create platforms that act as meeting points. Initiatives are taken that bring together resources within industry, education and research. Libraries, theatres and museums also find common forums. These are forms of co-operation that are characteristic of *network solutions*.

It has taken generations to create national identities. Although the spatial proportions would appear to be relatively modest, it is going to take considerable time to create a *common* identity on the Danish and Swedish sides of Øresund. Personal contact networks and common experience create lasting structures. They are passed on from generation to generation by upbringing and education. This inherited knowledge is stored in our libraries and archives as well as in the architecture of our cities. It takes decades to build functioning institutions and generations to create the sense of unity and social fabric that is the basis of all societies. Questions related to the propensity to change over time are highly relevant to the discussion below.

By the beginning of 2002, the discussion on the future of an integrated Øresund region had become concentrated on two issues. Firstly there was the question of the charges to be paid by motorists for the use of the fixed link. It is a widely held view that the present level of charges puts a brake on integration in the region. A zero charge would stimulate traffic and in the long run generate economic benefits to society. The other issue relates to differences between the Danish and Swedish taxation systems and the regulations governing insurance and social benefits. These differences are considered to make it more difficult for people to live and work on different sides of the Øresund. Cultural and linguistic questions have raised considerable interest in the region, particularly regarding the desirability of similar identities and shared mental maps among people who live and work in the Øresund region.

From barriers to opportunities

Both the debate within the region and most of the studies that have been carried out often start out from the premise that there is a latent need for mutual contacts on both sides of Øresund. Due to various obstacles and barriers, it has not been possible to satisfy this need. Resource allocation has not been optimal and development in the region has been held back. Consequently attention has been concentrated on the reduction of these barriers.

A rather different perspective will be adopted in the discussion below. This does not mean that we completely disclaim the prevailing view. However other aspects have also to be taken into account.

> It is our ambition to try to break the impasse which is likely to prevail from a concentration on issues related to obstacles and barriers. Instead we would wish to draw attention to some of the opportunities that exist. In contrast to earlier approaches, we will examine how the need for contacts may be created and stimulated. Are there measures and *policy instruments* that can be employed to mobilise productive forces and co-ordinate the use of resources?

It seems probable that it is a combination of a large number of specific initiatives that constitute a whole range of micro-processes which will *together* ultimately generate the forces for change and transformation.

These initiatives may be undertaken by decision-makers in firms, public authorities, mass media, cultural institutions, education, research and private households. Since the

It seems probable that it is a combination of a large number of specific initiatives that constitute a whole range of micro-processes which will together ultimately generate the forces for change and transformation.

majority of these initiatives are likely to be conducted on a small scale, they will not in themselves attract much attention.

The central concern in this report is to discuss the local and regional impacts of the localisation of research activities. The significant impetus that a research centre may provide towards regional integration in a cross border region is of special interest in this context. As was mentioned in the introduction, the *example* to be discussed here has been determined a priori.

However, the analysis and surveys to be presented in this report are not only relevant to this specific context. Other scientific and knowledge based activities could just as well have been used as prerequisites for our discussions.

3

EUROPEAN
SPALLATION
SOURCE

The ESS project – the European Spallation Source – is concerned with the location and development of a third generation neutron source somewhere in Europe. Similar sites for Neutron Spallation are under construction in the USA and Japan. The planned site for this advanced research project will place great demands on both material and intellectual resources.

The need for land is estimated to be 1.3 mill sq kilometres while the investment expenditure is expected to be in excess of 15 billion SEK. The personnel requirements are expected to be in the region of between 500 and 600 persons on a permanent round the clock basis. It is anticipated that as many as 4 000 to 5 000 scientists will make use of the plant every year.

The location of the research centre remains an open question. In Germany, France and the UK competing national and regional interests are prevalent. The existence of substantial competition between the major nations within the EU ought to provide the Scandinavian countries with certain opportunities. The question then arises as to where in Scandinavia the plant ought to be located ? As stated above, the aim of this report is to examine the case for locating the research centre in the Øresund region and to discuss its impact on the region. Finally the question may be raised regarding the best site for the plant within the region.

As stated above, the aim of this report is to examine the case for locating the research centre in the Øresund region and to discuss its impact on the region.

A *brief survey* of the ESS project is presented below. A more detailed scientific analysis carried out by experts in the field of nuclear physics is presented in Appendix to this report.[4]

The essence of matter

The study of what is termed "condensed matter" in disciplines such as physics, chemistry, material science and biology constitutes the scientific basis for a wide range of phenomena in modern societies. Increasingly greater demands are placed on the functional qualities of materials in a wide range of disciplines from information technology to building technology, electronics, chemistry and medicine. There is a growing need for a greater understanding of the basic structures and dynamic properties of materials at the atomic and molecular level. The science of condensed matter is being confronted with highly complex problems which represent a major challenge to experimental and theoretical research in this field. Although research into condensed matter is by its nature "small scale", researchers have a need for "large scale" research facilities for their work.

Why neutrons?

As early as 50 years ago, the Nobel prize winners, Shull and Brockhouse were able to demonstrate the unique qualities of neutrons which make them particularly suitable instruments for studies of "where atoms are and what atoms do". Among their advantages we find the following characteristics:

- neutrons have the mass of hydrogen

- neutrons are neutral, element-specific probes which interact with the atomic nucleus in matter rather than with the electrons surrounding the nucleus. Neutrons may thereby distinguish between neighbouring elements

- neutrons probe magnetism at the nanolevel because they have a magnetic moment that interacts with other magnetic moments in matter. This interaction is of the same order of magnitude as the interaction with the nuclei

- neutrons interact relatively strongly with light elements and differentiate between isotopes, in particular between hydrogen and deuterium. They act as an in vivo probe of hydrogen at the nanolevel

- thermal neutrons have wavelengths comparable to characteristic length scales in matter. Accordingly thermal neutrons are well suited to studies of the static and dynamic properties of most terrestrial phenomena

- neutrons are highly penetrating, weak coupling with absolute and readily interpretable cross-sections

- neutrons penetrate deep into matter. They are non-destructive and cause no damage.

What is spallation?

Neutron sources have been traditionally based on reactor technology. As is well known, this technology is associated with the problems of radioactive waste and reactor security. The ESS almost entirely eliminates radioactive waste since it is based on the spallation effects of efficient neutron production. Spallation occurs when a fast particle, the high energy protons, bombard a heavy atomic nucleus. This process which results in the production of a number of neutrons is called spallation. When the bombarded nucleus is heated up, further neutrons are generated. This process is comparable to throwing balls into a pail that already contains a number of balls. A number of them are ejected from the pail right away to be followed after a while by several others that are bouncing around. Each proton that collides with a nucleus "ejects" 20 to 30 neutrons.

Significance

Neutron scattering is an experimental technique that has been used for half a century to study the structure and dynamics of matter. The technology is however strictly limited by the intensity of the available neutron sources. Since the 1950s, this intensity has been able to be increased by a factor of four. In fact this increase was achieved already thirty years ago by the establishment of the world's leading neutron research centre at the *Institute Laue Langevin in Grenoble.*

The scientific and technological problems have subsequently increased in complexity, subtlety and scale. There is an urgent need to develop a third generation neutron source. Compared to existing neutron sources, the proposed ESS plant would increase the intensity of the available neutron sources by a factor of between 10 and 100. This would be the greatest single advance in terms of performance since the 1950s.

The implications for research will be of major importance. The role of the ESS in relation to research into condensed matter will in some respects be comparable to the impact of the Hubble Space Telescope in astronomy. The Hubble telescope has changed our conception of outer space by giving us the opportunity to see further with greater clarity than ever before; phenomena that were on the outer limit of what was observable together with completely new phenomena are now within range. Analogously, the new scientific phenomena and technological functions that will be revealed as a result of the ESS studies into condensed matter will create *a new inner universe.*

Moreover it is anticipated that the efficiency and importance of the ESS will become an increasingly vital complement to the existing, less powerful neutron sources. Once again a parallel may be drawn with the Hubble telescope that has become an integrated part of a network of less powerful land-based observatories.

A group of about 70 researchers from different fields of research met in Engelberg in May 2001 to discuss the major areas where the ESS centre would play a significant role. The object of the meeting was to discuss the optimal parameters for the ESS and to identify the key areas that are at the cutting edge of what may be achieved at present. The meeting resulted in a list of major challenges in solid state physics, material science, biology and biotechnology, chemistry, geology, soft condensed matter and neutron physics. It was concluded inter alia that the ESS would create the following opportunities:

1 Allow researchers to come into contact with new problems and to pose new questions

2 Provide new means to tackle problems at the research frontier

3 To offer high quality experimental data in support of unambiguous conclusions regarding theoretical models

Some of the expectations held by researchers regarding the planned ESS centre are presented in the *Appendix*. Among the highly challenging special areas mentioned by the scientists are the following:

- **SOLID STATE PHYSICS**

- **MATERIAL SCIENCE AND ENGINEERING**

- **BIOLOGY AND BIOTECHNOLOGY**

- **SOFT CONDENSED MATTER**

E g polymers, floating crystals, micellar solutions, micro-emulsions and various biological materials. These substances have a wide range of uses such as structural and packaging materials, foam and plaster, cleansing substances and cosmetics, paints, foodstuff additives, lubricants and fuel components, rubber etc.

- **CHEMICAL STRUCTURES, KINETICS AND DYNAMICS**

- **EARTH SCIENCE, ENVIRONMENTAL SCIENCE AND CULTURAL HERITAGE**

Modern spallation sources allow us for example to carry out detailed studies of the crystal structures of minerals. It becomes possible to model fundamental processes on the planet from large scale phenomena such as major earthquakes and volcanic activities, transportation of pollution into the earth's crust and the conservation of stone in buildings and monuments.

● **LIQUIDS AND GLASS MATERIAL**

Amorphous material play a central role in our daily lives. Two thirds of the earth's surface is covered by water. Most of our body is filled with the substance. Glass material is to be found in windows, optical fibre and even confectionery. It is also used to provide a stable outer surface for medicines. Ionic conductors are found in car batteries (electrical cars in the future), mobile telephones and computers. Despite their widespread use, our knowledge of these materials is highly limited, particularly in relation to crystalline materials.

● **BASIC PHYSICS**

Our image of the world has changed dramatically during the past 25 years, from the building blocks of elementary particles to the structure of the universe. Neutron physics has made important contributions in this area. At the universal level, cosmology has developed into an exact science while neutron physics has provided us with insights into the creation of matter and the various transitional phases in the history of the universe. Different data extracted from measurements of neutron beta decays has shown that there are three basic families of particles. At a much smaller level, neutron experiments have also provided us with a new understanding of strong, electron-weak and gravitational interaction. Neutron interferometry and neutron spin-echo experiments have shown how non-classical states may be created and used for high-sensitive studies of condensed material and basic physics research. Within these fields of research in basic physics, there are a range of front-line experiments that are waiting to be carried out in a new advanced ESS establishment.

4

LOCAL AND REGIONAL IMPACTS

The question of the socio-economic impacts, local and regional, of the location of new plants or the growth of existing production facilities have been of central interest in regional policy in European countries for a number of decades. This may be seen in the discussions and reports on the consequences of both the establishment and closure of industries and military regiments as well as in the relocation of central government offices. In recent years, the location of new universities and university colleges has attracted the greatest attention at the national, regional and local level.

The approaches and theoretical perspectives adopted in various academic disciplines during the 1950s and 1960s were characterised by a well founded belief that policy in the areas of economy, research, higher education and labour market should be largely conducted within national frameworks. Within the decision-making domain of the territorial state, both places and regions were not considered to be to any great extent *directly* dependent on conditions outside the boundaries of the country. In the few cases where this type of dependence could create problems, it was the duty of the state to intervene. During the past two decades, the growing direct cross-border dependences of places and regions has become increasingly evident. At the same time, the content of economic and working life has been subject to drastic change. Hence it is hardly surprising that the approaches of social science researchers to social and economic problems has changed. At the same time, the character and content of impact analysis and effect models have been altered.

Multiplier effects and cumulative causation

Traditional effect models are constructed in the following manner. Large investments in plant and machinery create a local and a regional demand for goods and services. The expansion of new and existing workplaces create a larger labour market. The growth of employment together with family dependants leads to an increase in the population. The growth of purchasing power expands the retail trade and the market for household services in the area which in turn generates higher tax revenues for the local authorities. These effects may be considered to be *direct* and are relatively easy to calculate.

The *indirect* effects are not as immediate and are much more difficult to determine and assess. Employment is not just created in new or expanding workplaces. There is a whole range of sub-contractors and service firms that also are able to increase their production as a result of the expansion of demand. Public authorities also find it possible to increase their investments and expenditure which generates further expansion. It should be borne in mind that these multiplier effects do not only operate in a positive direction. Plant closures and the rationalisation of production units lead to consequences that operate in the opposite direction, albeit usually subject to a certain time lag.

There is considerable evidence to show that these effects have their greatest impact on places that have a small one-dimensional labour market and a highly restricted range of services. Large, diversified regions with dense supplier and customer networks will be least affected since there is usually a degree of more or less latent over-capacity among sub-contractors, retail trade outlets and service companies. Large regions are also more robust in relation to closures and layoffs since they are able to offer a wider range of employment alternatives than is possible in small. According to traditional studies, a multiplier of between 2-3 could be expected in larger regions i e 1 000 new employment opportunities could raise the population in the area by between 2 000-3 000 persons. In smaller places with limited labour markets, the employment multiplier could be as large as 5-6.[5]

A well-known example from the extensive literature in this field is the description by the Swedish economist, Gunnar Myrdal, of economic progress and decline as a process of *cumulative causation* that is more or less self-generating and difficult to influence once it has started. Positive growth processes tend to become concentrated in a few central areas of expansion while backwash effects characterise many peripheral regions. Selective migration, capital movements and free trade favour the areas undergoing rapid growth while adversely affecting the areas experiencing slower growth. The free play of market forces tends to increase rather than reduce regional economic inequalities.

These backwash effects may be offset by positive spread effects which would appear to be greater in countries that have attained higher levels of economic development. This is largely attributable to the fact that favourable economic conditions tend to generate improvements in communications, higher levels of education and an increased prepared-ness to remove obstacles to the spread of welfare.[6]

Numerous research reports have over the years followed the theoretical approach adopted by Francois Perroux who coined the concept of "pôle de croissance". Initially this concept referred to sectors or groups of firms within an economy which were strongly linked together. Growth in this type of sector could provide substantial spread effects. As the French term implies, the growth pole concept may be viewed as a crossing point between sectors (rows and columns) in an input-output table.

In relation to the analysis below, it is of particular interest to note that Perroux viewed growth not just as a quantitative process but as a qualitative process characterised by transformation and renewal. The innovation processes (development of new products, new technology and new production processes) and the diffusion of innovations are seen as important driving forces in the economy.[7]

Here there is a need en passant to remind ourselves of the pioneering work of the Austrian economist, Joseph A. Schumpeter, on how innovations create business cycles of growth and depression in a sequential order. By means of "creative destruction", development moves forward in waves in a capitalistic economy.[8]

Numerous authors among economists and geographers have given the concept of growth pole a purely geographic meaning. Growth poles are places where different types of growth become concentrated, either spontaneously or as a result of planning. These ideas were especially prevalent in the regional policy debate of the 1970s and helped to form many of the growth related policy measures that were undertaken in the Nordic countries and in other parts of Europe.

The need for a wider perspective

With access to specific data on for example an EES establishment of the aforementioned type – buildings, equipment, personnel, rates of utilisation by visiting researchers – it is entirely feasible to conduct an analysis of the impacts of the project. However, several factors would appear to suggest that we are confronted with a complex of issues that need to be analysed step by step in future research.

The most immediate central question that arises is whether or not a major research establishment can be expected to play a more strategic role in its environment than for instance a regiment or a factory. The few thorough studies that have been carried out into the effects of newly established universities indicate that this is the case although the impact is not as great as the public debate would appear to suggest.

This issue was examined by the Dutch economist, Raymond Florax, in his book *The University - A Regional Booster*. Here he analyses the extent to which universities may be expected to contribute towards radical changes and considerable development in a region.[9]

New universities unleash a wave of building activity. This can be seen throughout Europe. A university with its students, teachers and researchers form one of the largest workplaces in the region. Highly educated and well paid people create a particular type of demand for goods and services. The continuous turnover of young people creates a special dynamism in the region. A pool of well educated labour gathers around the university, people who are attractive to both private and public employers. From a Swedish perspective, these types of effects are easily observed in towns such as Lund and Umeå.

In many university towns, there is a strikingly large element of established cultural institutions and recurrent events. Computer and IT consultants, hi-tech firms and other specialist enterprises are attracted to university towns. Relative to their size, they also attract a considerable number of publishers and printers. In Lund for example, a town of around 100 000 inhabitants, there are more than one hundred publishers and a large number of printing businesses. The extent to which universities are both an expression of and an actor within the physical and cultural infrastructure is a question that has not previously received much attention. Currently it attracts considerable attention as part of an ongoing research programme.[10]

The second question is more difficult to examine in a strict research perspective but ought nevertheless to be raised. It concerns the *psychological impacts* of major physical infrastructure projects. It was discussed earlier in relation to the symbolic value of the fixed link between Sweden and Denmark. It became evident both during and after its construction that a concrete, highly visible building structure both attracts interest and acts as a source of inspiration. The bridge has a major part to play in putting the Øresund region on the world map.

As will be discussed below, research is currently surrounded by something of a halo, especially in the fields of science, medicine and technology. Places that are associated with successful research become renown as for example in the cases of CERN in Geneva – the European centre for particle physics – and the Institute Laue Langvin in Grenoble mentioned above.

With ESS in the centre

Figure 4 provides a simple illustration of how an ESS plant may form the centre of a complex of activities. Around the central core, there is a ring of scientific spheres that are more or less dependent on the activities that are conducted at the neutron spallation source. (Compare the presentation in the previous section and in the *Appendix*). The figure should not be interpreted as a map but as an illustration of how different activities are dependent on one another. The extent to which these dependencies exert an influence on the geographical location is a question that will be examined later.

Outside the ring of scientific spheres, there are several examples of activities and industries that can be considered to be more or less directly or indirectly related to these spheres. In the majority of cases, the research findings are not likely to be directly applicable within industrial sectors

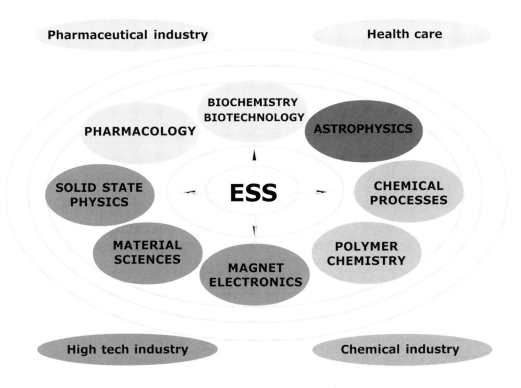

Figure 4   Research areas and industrial cluster formations around ESS. Outline

without applied research and development of new products and processes. The extent to which this research is able to find practical outlets is likely to increase as one moves from the centre to the periphery of the figure.

It is appropriate at this juncture to provide a brief definition of the innovation systems and technological clusters. These concepts are of special interest when we are going to discuss the need of geographical proximity in a knowledge based economy and the future of the Øresund region.

Innovations and innovation systems

Our view of industrial location and economic competitiveness is based on the idea that a firm's long term prospects are determined by its capacity to innovate. Naturally cost advantages are of importance. In the long run however, it is the ability to generate and utilise new knowledge which makes it possible to produce better products and employ more efficient manufacturing processes which in turn create the preconditions for survival and development.

Technological innovations apply to an entire range of processes from invention to the marketing of the finished product. What is termed product innovations may comprise new products or new variants of existing products. Process innovations lead to productivity gains as a result of a more rapid production process, new machines or new forms of organisation that allow the goods to be produced more efficiently and at lower cost.

The innovation concept does not have to be limited to the production of goods. It may also be applicable to new forms of service, distribution and administrative routines within both the private and public sectors. In recent years, the question of the relationship between institutional conditions and innovation processes has received particular attention.[11]

Isolated firms are seldom innovative. The majority of firms modernise their production process and develop new products and services by interacting with other firms and institutions – customers, competitors and suppliers of inputs and services. Laws and regulations together with different cultural patterns form the framework for this co-operation and interaction.

Innovation processes in networks of relations may take considerable time to develop and they may be difficult to follow in detail. It is a question of evolution rather than revolution when research findings and inventions are developed into successful commercial products.

A linear model has previously been used to describe the role of university research in innovation processes. Here the chain of causality is simple and straightforward. Basic research is the first step. This type of research, especially in Sweden, is mainly conducted in universities and institutes of technology financed by government and special research foundations. Applied research involves the further development of the advances made by basic research. This may sometimes be conducted in university departments or increasingly, in the privately financed laboratories of firms and special research institutes. In a final stage, the firm develops new products and processes that are commercially viable and provide new employment opportunities and ultimately increased welfare.

In certain contexts, this linear model is still relevant. However it is limited in the sense that it provides an excessively schematic and simplified view of the complicated relations that often prevail in a knowledge based economy. More complex models have recently been developed in order to increase our understanding of the ways in which research may affect economies and welfare. Innovation processes are assumed to take place within a context of *interactive learning* involving a number of different actors. *Innovation system* is one of the names for the structures that have replaced the simple chain of causation provided by the linear model. The term system refers to a network that binds together institutions or actors that have mutual contacts and trust. This applies for instance to university researchers, decision-makers within public administration and various types of firm. Concepts such as *technological system* and *technological cluster* are used in relation to analyses of predominantly technical innovation processes.

Studies of innovation systems and technological clusters are characterised by different approaches and perspectives dependent on the types of question that have been raised. Studies of the Swedish export industry represent almost a *sector perspective*. Here special studies have been devoted to transport related, wood products related, foodstuffs or pharmaceutical related industries. However, as is the case with many older studies, a *national perspective* is also present which indicates that networks and clusters are assumed to possess special national characteristics. Within the boundaries of territorial states, the legal and regulatory systems and the policy measures of central government are applied in a uniform manner. Within these boundaries there are also cultural patterns and institutions that have different characteristics and functions than elsewhere.[12]

As mentioned in the second chapter of this book, the Danish and Swedish systems are remarkably closed in relation to each other. This is presumably one of the major hindrances to rapid economic and cultural integration between the Swedish and Danish parts of the Øresund region.[9] Nevertheless it is undoubtedly the case that over time, internationalisation will lead to an acceleration of the innovation processes that cut across national boundaries e g within global companies or by means of technical co-operation between firms. Undoubtedly the establishment of a major research centre such as ESS in the right location will provide a marked impetus to this process.

Clusters

As will be pointed out below, network is a concept on which there is considerable contemporary focus. The same could be said of the concept of cluster. This leads in turn to another related concept namely development bloc. It was above all through the work of the American economist, Michael Porter, that the cluster concept attracted considerable attention following the publication of his major work in 1990.[10] Porter argued that it wasn't actually nations that competed with one another but rather firms. He then went on to use the cluster concept to describe how international competitiveness is created within a group of related firms.

The formation of clusters does not need to imply that the activities concerned are bound together by a close physical proximity. It is quite sufficient at least at the outset that the firms are part of a coherent, functioning system. Here it is appropriate to remind ourselves that Francois Perroux's concept of "pôle de croissance" referred primarily to sectors or groups of firms within an economy that have strong links to each other and that growth in this type of sector produced strong spread effects throughout the entire economy.

In recent years, it has become more common to refer to clusters within a specific geographical environment where related firms are localised and surrounded by supporting activities. The various actors – all of whom are linked together - may be subcontractors, customers, competitors, universities, authorities and organisations. There are numerous examples of geographical clustering such as Hollywood in relation to the film and entertainment industry, Silicon Valley within information technology, Detroit in relation to motor vehicles and the City of London regarding financial services.[15]

> **In recent years, it has become more common to refer to clusters within a specific geographical environment where related firms are localised and surrounded by supporting activities. The various actors – all of whom are linked together - may be subcontractors, customers, competitors, universities, authorities and organisations.**

The arguments above brings us closer to the approaches and research findings associated with "The New Economic Geography" (p 90). As the British geographer Ron Martin points out, it is really only the concept of cluster that can be said to be new. The underlying arguments and observations have been available in the economic literature for many years.[16]

Elements of a continuous debate

Before proceeding with an analysis of the socio-economic impacts of the research centre, it is appropriate to examine some of the components that are part of what has become an increasingly complex effect model. At the same time it is essential to extend the socio-economic and geographical perspective.

It is above all two factors that make it vital to integrate the plans for the location of a future research establishment within a wider perspective. Firstly a research centre of the ESS type takes on a strategic importance in what is termed a knowledge based economy. Secondly the establishment of this research centre should not be seen as a local or regional concern only. It has a much wider geographical scope that stretches beyond the boundaries of Denmark and Sweden. In our discussions the European perspective will be essential.

The components below deserve a closer analysis. Conditions and development characteristics that can be expected to apply at a general level are discussed under six separate headings. Conditions that are more uniquely related to the Øresund region are discussed under the last three headings. The central idea of a world undergoing rapid change will be emphasised throughout. As was mentioned in the introduction, the presentation here should be seen as a series of empirically based sketches that draw on the relevant literature in order to provide a basis for an ongoing discussion and future research.

- **THE KNOWLEDGE-BASED ECONOMY**

- **THE GEOPOLITICAL PERSPECTIVE**

- **A EUROPE OF REGIONS**

- **THE NEW ECONOMIC GEOGRAPHY**

- **THE MECHANISMS OF REGIONAL SUCCESS**

- **CULTURES OF CREATIVITY**

- **THE ØRESUND REGION IN THE EUROPEAN URBAN LANDSCAPE**

- **INDUSTRY AND RESEARCH IN THE ØRESUND REGION**

- **COMPLEMENTARITIES IN THE ØRESUND REGION**

5

The
KNOWLEDGE-BASED
Economy

The industrialised world has over the past three decades undergone a process of radical changes. This transformation is so far-reaching that researchers and political commentators have felt the need to discuss these changes in terms of a new era in human civilisation. Whereas previous periods of history have been characterised by a relatively slow transition from agrarian to industrial forms of production and living conditions, the present transformation is unique in human history in terms of the pace at which technology, forms of production and economic, social political and cultural conditions have changed.

A range of terms has been used to denote this era. Three examples have been chosen in order to describe how a new period of history is taking on sharper contours and a richer content at the same time as we gain a new perspective on the conditions for economic and social welfare.[17]

The Post Industrial Society

Historians have identified at least two industrial revolutions. The first began in England at the end of the eighteenth century under the impetus of technical inventions such as the steam engine, the mechanical loom and metallurgical processing. It was characterised by the replacement of hand tools by machines. The second industrial revolution began in the mid-nineteenth century and is associated with improvements in the manufacture of steel, development of electricity and the introduction of the combustion engine, new types of chemicals and the telegraph and telephone. This second industrial revolution was based on what we would now call scientific research whereas the first industrial revolution was first and foremost dependent on advances in craftsmanship and engineering.

As Manuel Castells, the Spanish-American sociologist, has emphasised, the transition from an agrarian to an industrial society was undoubtedly characterised by revolutionary changes. Unexpected waves of technological innovations and new applications transformed production and distribution processes, created an abundance of new products and led to a radical shift in the geographical distribution of wealth and power. The rise of Western society which was in practice limited to the United Kingdom and a handful of countries in Western Europe together with their North American and Australian offshoots was largely based on the technological superiority that was achieved during these two industrial revolutions. To a certain extent, the second industrial revolution that was more heavily dependent on scientific progress than the first brought about a shift in industrial power to Germany and the United States where the most important advances in chemistry, electricity and telephony had taken place.[2] Peter Hall and Pascal Preston have stressed the importance of local factors underlying the geographical redistribution of technical innovations after the mid-nineteenth century. Berlin, New York and Boston became the world centres of advanced technology between 1880 and 1914 while London became almost a faint copy of Berlin during this period.[19]

Historical experience shows that technological breakthroughs often occur in clusters both in time and space. Technical innovation has reflected a certain level of knowledge in a given place or area. It has been dependent on a specific institutional and industrial environment. The closer the relations between technical inventions, innovations and practical applications, the more rapid the rate of economic and social transformation and the greater the benefit to society in terms of improved economic and social conditions.

According to many analysts, the advanced industrial countries are now undergoing a process of revolutionary transformation. In 1969, Alain Touraine, published his work, *La Société post-industrielle*. In this and subsequent works, he showed how material goods were being replaced by services to form a new core of production in the post industrial society. This discussion was widened by Daniel Bell who analysed the radical effects of new technology on industrial processes in terms of their drastic impact on occupational and social structure.[20]

The transition between different forms of society takes on a range of forms in contemporary society. Trade between industrialised countries has changed character. A growing proportion of the exchange of goods comprises high value-added goods based on inputs of advanced skills and technology. The service component has also increased. This has led to a corresponding decline in the value of the pure material content of production. New pharmaceutical and electronic products provide good examples of industrial processes where the value of raw material inputs has diminished to an almost minimal level whereas a large investment in research and development has been required in order to increase the attractiveness of these products on the market.

The composition of production and employment has also undergone radical changes. This is borne out by the available statistics for the OECD countries on national income, population and industrial output. Since the 1950s and 1960s, the GDP and employment shares of raw-material based industrial production has declined in all OECD countries.

A growing proportion of the exchange of goods comprises high value-added goods based on inputs of advanced skills and technology. The service component has also increased. This has led to a corresponding decline in the value of the pure material content of production.

The proportion of services in output and employment has correspondingly increased. Before 1990, the occupational share of agriculture, mining, construction and manufacturing was already under 30 per cent in Sweden, West Germany, USA and Japan. Service employment comprised more than 70 per cent of total employment. The term "service" includes transportation, communications and business and household services in both the private and public sectors.

The figures presented above represent only the tip of an iceberg of fundamental structural changes in the economy and labour market. A closer examination of the manufacturing sector viewed by many as the engine of the industrialised economies may serve as an example. Even ten years ago, the share of wage costs that was attributable to various forms of information processing was as much as 60 to 70 per cent. Company in-service training, co-ordination of firm operations, internal communications and the external purchase of information services all fall within this area.

Research and development was at that time the sector undergoing the most rapid growth. The proportion of qualified, highly educated staff became an increasingly important factor in production. For example in the USA, about thirty per cent of working time in manufacturing firms was devoted to assignments that required post-secondary, academic education. Given the salaries prevailing at that time in the USA, this meant that 25 per cent of the labour force produced around 75 per cent of the value of production.[21]

The Information Age

The strategic role played by the processing and communication of information in present day society has received particular emphasis in the work of Manuel Castells. He draws an important distinction between the concepts of *Information Society* and *Informational Society*. The former emphasises the major role played by information in the development of society. This is a fairly banal, self-evident observation. In its widest sense, information communicates knowledge which has been of decisive importance for the majority of societies throughout history. On the other hand, the concept of informational society points to a specific form of social organisation where the collecting, processing and distribution of information under the influence of new technological conditions becomes a basic precondition for productivity, economic and social development and power. The concept of the network has also taken on a strategic importance in discussions of the growth of the informational society. As will be seen below, these concepts of network and informational society are actually important complements to each other.

Castells depicts the modern city as informational. New information technology, an increased demand for knowledge and the growing importance of information processing has brought about a distinct segmentation of labour particularly in major cities.

New, highly paid, employment opportunities are created in sectors such as advanced business services, high technology manufacturing,

design, entertainment and within mass media. At the same time however, there has also been an increase in low paid service jobs in hotels, restaurants and household services. On the other hand, the number of employees in traditional manufacturing jobs that previously had formed an intermediate level in many cities has declined drastically. This has created a Dual City, a city characterised by segregation between high income earners and the low-paid, the highly educated and those who lack a basic education.

Many of the latter are to be found among immigrants who belong to ethnic minorities and work in the growing informal sectors of the economy.[22]

In three major works gathered together under the unified title of *The Information Age, Economy, Society and Culture*, Manuell Castells has provided a penetrating analysis of the growth of the informational society. A technological revolution centred on advances made in information technology has brought about rapid changes in the foundations of society. Microelectronics, computers (hardware and software), telecommunications and the broadcasting media are all encompassed within this technological complex. The invention of the transistor (1947), integrated circuit (1957) and microprocessor (1971) were notable milestones. Castells also includes genetic manipulations as a new form of information technology.

The core of this transformation is the processing and communication of information. Around this central area, we have seen a series of major technological breakthroughs during the final two decades of the twentieth century in fields such as material technology, energy supply, applied medicine and manufacturing and transportation technology.

The economy has become global. Production, distribution and consumption of goods and services are now organised on a global scale – either directly or through networks of contacts between economic agents. Castells has coined the expression: "From a space of location to a space of flows." It is no longer a matter of analysing the location of physical plants. It has become much more important to study the flows and the interdependencies which bring together units that may or may not be in close proximity to one another. The most important parts of this knowledge based production lies in the linkages and flows of the networks.

In the future, the international economy will most likely create varied geometric patterns of consumption, work, capital and management.

Production will to an increasing extent become dependent on forces outside a particular location and thereby increasingly influenced by networks that communicate signals and resources. It goes without saying that human beings will continue to live in particular places and be dependent on specific neighbourhoods. However production and power will not as is presently the case be identifiable as points on a surface or as part of a defined area. Instead they will become integrated into networks. Social reproduction will remain specifically local while production becomes organised in a geographic space of flows. In this new set of relationships, the territorial state will become fairly power-less.[23]

The Knowledge-Based Economy

The development of knowledge is an inherent part of the process that we have described above. Naturally all technical progress implies new knowledge. However the emphasis was previously on the use of knowledge and skills to convert raw materials and semi-manufactures into finished products. What has emerged now is that information has become *per se* the most important input and at the same time, the new product. The strategic machinery is no longer that which processes the material but rather that which facilitates the information processing and the control and steering of production processes.

The current changes that are now penetrating and transforming the material base of the economy, the social and cultural life are comparable to those generated by the great industrial revolutions of the eighteenth and nineteenth centuries. A close tie has been established between the social processes that govern the manipulation and creation of symbols (important cultural attributes) and the capacity to produce and distribute goods and services (productive forces). *For the first time in history, the human intellect is a direct productive force, not just a vital element of the production system.*

As mentioned previously, research and development work accounts for an increasingly large share of the value added in production and trade. This is one of the signs of a growing knowledge based economy. The expansion of formal education and the wide respect shown to science and research provide us with other indications. It is not just a question of a growth in the volume of knowledge based on qualified research. This knowledge is also finding its way to parts of society that were largely unaffected by the work of researchers.

Research has actually become our fastest growing industry. For example between 1990 and 1996, the total number of workplaces in Sweden declined by 3.6 per cent and the total number of employees by 12.6 per cent. This was equivalent to almost 300 000 employment opportunities. This fall in employment was particularly evident during the period 1990 – 1993. However as may be seen from Table 1 below, there were substantial differences between the various sectors of the Swedish economy.

The severe contraction in employment opportunities was above all concentrated in the manufacturing sector. In the table, the sector denoted as "others" that includes agriculture, trade, construction, transport and certain other services also experienced a marked decline in employment. The rapid decline in the manufacturing sector occurred in the early 1990s. A slight recovery in industrial employment has subsequently

Table 1 Relative changes in the number of workplaces and employees by sector, 1990-1996

Sector	Number of workplaces	Number of employees
Manufacturing industry	-12.6	-16.7
Professional services	12.4	0.9
Research outside the industrial sector	49.5	49.7
Others	-6.6	-15.9
All sectors	-3.6	-12.6

Source: Statistics Sweden

taken place (see Table 6, p 139). On the other hand, professional services have recorded considerable growth - composing e g technology trading firms, financial services, IT consultants, technological consultants and advertising firms. *The single greatest area of growth was in the field of research*, including both state financed university research and privately financed research outside the manufacturing sector. In the table, it should be noted that the research-intensive parts of the manufacturing sector are included within the industrial sector. However it will be treated separately below.[24]

The increase in the time spent in formal education is one of the most prominent characteristics in the growth of industrialised countries. In less than one hundred years Sweden developed from being one of Europe's poorest to one of the continent's rich-

The increase in the time spent in formal education is one of the most prominent characteristics in the growth of industrialised countries.

est countries. During this period, the average length of time spent in formal education increased from three to eleven years per employee. The enormous expansion in higher education occurred in the post-war period.

As late as the 1930s and 1940s, less than one per cent of an age cohort received higher education. In the late 1990s, the corresponding figure was 40 per cent. At the end of the Second World War, there were less than 40 000 university graduates in Swedish society as a whole. Today the corresponding figure is about one million. The number of post-graduates has also increased but not to nearly the same extent. The elite universities have become the mass universities. From the late nineteenth century until the 1960s, Uppsala, Stockholm, Gothenburg and Lund were the exclusive centres of higher education and research in Sweden. Today there are more than 30 higher education institutions in the country. This figure could be further raised by taking account of all the various centres for distance and specialised education.

Since the nineteenth century or in some cases earlier, university research and higher education have been largely financed and regulated within a national framework. In a country like Sweden, the established church, educational system, health care and public administration have been the principal sources of employment for university graduates. The career paths open to graduates have been largely confined within the country's own borders. As Nathan Rosenberg has clearly shown, the occupational demarcation between science and industry was almost total in the Western world right up until the end of the nineteenth century. Even after this date and despite certain common sources, science and industrial technology have been largely developed along different paths that are easy to separate and seldom cross. It was not until far into the twentieth century that innovation systems were developed in certain areas that indirectly connected universities and firms. It was in the industrial research laboratories that the recruitment of university graduates began to bridge the gulf between science and industry.[25]

However the major breakthrough occurred with the outbreak of the Second World War. A large part of the scientific and technical capacity of the warring states was devoted to total warfare. The Manhattan project in the United States which was devoted to the manufacture of an atom bomb was for its time

However the major breakthrough occurred with the outbreak of the Second World War.

the most spectacular example of close co-operation between the war industry and science. Competencies from a range of ethnic backgrounds were gathered together in secret laboratories. Research that would normally have taken decades in peacetime was carried out in a couple of years. Entire industrial towns were built in a hurry to house applied research and development facilities. Never before had such large sums been invested in research.

In a similar fashion but under somewhat freer conditions, a gigantic space research programme was launched. During the days of the cold war, co-operation between research and industry was further developed, not just in the USA but in the UK, France and other countries as well. The strongest ties between university research and industry would still appear to be within what is known as the military-industry complex. The foodstuffs and pharmaceutical industries are examples of other similar blocs of co-operation.

As a result of these developments, the practical significance for the nation of academic research and particularly scientific research became increasingly evident. The status and prestige of scientists grew in an unprecedented fashion. The increasing stature of research spread from science and medicine to encompass all types of research. At present there would appear to be a widespread view that the university is one of the major driving forces behind technical and industrial development. There is also considerable discussion of the idea that there is a fundamental relationship between research and education in general and the international competitiveness of firms - and consequently, full employment and welfare.

Traditionally, science and the production of knowledge has been associated with universities and specialised research institutes. The work has been often conducted in hierarchical organisational forms within fairly closed disciplines and generally subject to a strictly regulated quality control. Many have considered the autonomy of the universities to be an invaluable asset since scientific knowledge has been able to develop in a manner relatively free from the economic interests of the surrounding society.

In our contemporary society, it is no longer possible to make a strict distinction between what is termed science and research on the one hand and the rest of society on the other. Science, politics, economics and culture have become interwoven and interact with one another in a highly complex fashion. Different sectors have become so internally heterogeneous and externally dependent on one another that it is not longer possible to define them as clearly separate identities.

Scientific knowledge is no longer concentrated to certain places and sectors in society but has become "socially distributed". It is actually not particularly meaningful to treat science as something specific. It is no longer the case that it is just researchers at university departments and research institutes that analyse, estimate, theorise and use abstract models. These functions are increasingly carried out in different parts of society although the persons whom we call scientists are often more specialised and subject to a more thorough scientific control.

A number of factors have contributed to the dissemination and integration of research. As has been previously mentioned, elite universities have become mass universities. In certain countries, more than half of the youngsters in a particular age cohort may be enrolled in higher education. Graduates are increasingly found in a wider range of posts in a growing number of sectors.

Researchers are to be found in growing numbers in private firms and public administration rather than just universities and research departments. Independent experts, consultants and special teams of experts help to mediate knowledge and skills between traditional university research and various sectors of the economy. Informal contacts between researchers, experts and decision makers create complex networks that communicate knowledge and ideas. In conclusion it may be said that the production of knowledge in modern society has the following characteristics:

1 An increase in the number of potential sites where knowledge can be created; there is now an interaction between not just universities and colleges, but non-university institutes, research centres, government agencies, industrial laboratories, think-tanks, consultants.

2 The linking together of sites in a variety of ways – electronically, organisationally, socially, informally – through functioning networks of communication.

3 The simultaneous differentiation, at these sites, of fields and areas of study into finer and finer specialities. The recombination and reconfiguration of these sub-fields form the bases for new forms of useful knowledge. Over time, knowledge production moves increasingly away from traditional disciplinary activity into new societal contexts.

There has been a remarkable coincidence between the development of a more open system of knowledge production on the one hand and on the other the growth of complexity – and the increase of uncertainty in both. The climax of high modernity with its unshakeable belief in planning and predictability is now passed. Coincidence between the regularity of society and the predictability of a progressive science, was destroyed by two events. The first was the oil crisis of 1973-74. The second was the unexpected collapse of the Communism regimes in Eastern Europe and the end of the Cold War.

6

The Geopolitical Perspective

In the light of the transformation in society described above, our view of the preconditions and obstacles to national welfare and economic development has changed. The future of the industrialised countries no longer depends to the same extent as in former days on natural resources and the diligence of inhabitants. The development and communication of knowledge together with a capacity to innovate have become vital factors in promoting economic and social change. The role of science and the practical importance of higher education has become increasingly evident. The universities can be considered to act as a driving force for technological and industrial development. As has been seen above, this perspective on the transformation of society achieved a breakthrough during and after the Second World War.

The major breakthrough

Geopolitical developments during the 1930s and 40s helped to accelerate not just the scientific breakthrough but also brought about a change in the geographic distribution of the production of knowledge and economic power.

The United States emerged from the Second World War in a very strong position. On the other hand, its allies and the axis powers needed at least a couple of decades to recover. As discussed above in relation to the Manhattan project, the military sector was able to mobilise vast resources. This expansion of American industry was given a further stimulus by the Marshall Plan to aid rebuilding in Europe, the space programme and not least the rearmament programme in the shadow of the Cold War.

Of perhaps even greater importance to the establishment of American predominance was the human resources that flooded the country. Approximately 400 000 people were forced or chose to leave Germany or the areas that the German army had occupied in the 1930s and early 1940s. Among the émigrés were thousands of leading scientists, film-makers, actors, authors and artists, many of them of Jewish background.

Central parts of Europe were stripped of their intellectuals. Skills and knowledge became concentrated in certain parts of the United States and made a notable contribution to the country's remarkable technical, scientific and cultural development after the war.

A database of nearly 700 biographies of Nobel Prize winners clearly shows how unique knowledge and expertise were transferred across the Atlantic. Many Noble Prize winners received their prize after being appointed to prestigious positions at universities and research institutes in the US. These scientists grew up in Europe, were educated in European schools and universities, and worked in European research environments where pioneering discovers were first made. Many well-known winners of the Nobel Prize in physics, for example, converged in Princeton. Of the over 40 economists from different parts of the world that have won the Bank of Sweden's Prize in Economic Sciences in Memory of Alfred Nobel since 1969, 16 have been active at the University of Chicago.[26]

EU and US – two comparable magnitudes

Regional development, especially the prospects for a future Øresund region, is a focal point of interest in this study. Before proceeding with the discussion, it is necessary to examine the scale of this regional project and to demonstrate the highly confined framework within which we are operating seen in an international perspective. There are several reasons for choosing the EU and the US as measuring rods and objects of comparison. Scandinavian universities and firms have their major interests in Europe and the United States. Most of the literature that we have used in this report has been written by scholars at European and American universities. The colleges that are part of the aforementioned international research programme are also located in Europe and the United States.

As the European Union has developed, European countries have regained several of the leading positions in economy and research that they held prior to the Second World War. The present fifteen countries of the European Union have a combined economic and scientific potential that is comparable with that of the United States in the mid 1990s. Roughly speaking, the United States has 70 per cent of the population of Europe and slightly more than 80 per cent of its GDP. The United States has 29 per cent of world industrial production in terms of value added compared to 28 per cent for the EU. South East Asia has also a sizeable industrial production.

As the European Union has developed, European countries have regained several of the leading positions in economy and research that they held prior to the Second World War. The present fifteen countries of the European Union have a combined economic and scientific potential that is comparable with that of the United States in the mid 1990s.

Figure 5 shows the number of registered students in the United States, distributed across 250 urban regions ie cities and their Areas of Dominant

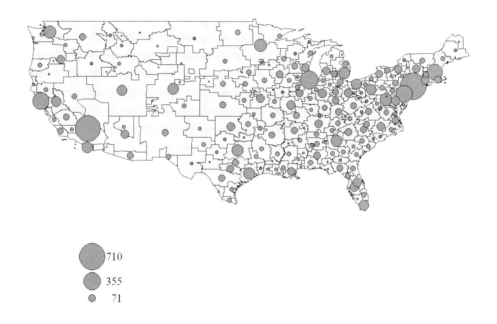

710
355
71

Figure 5 Students distributed by regions 1996. Thousands

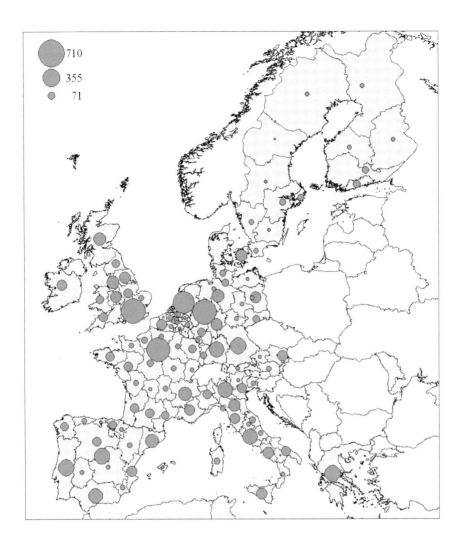

Figure 6 Students distributed by regions 1996. Thousands

Influence (ADI), a type of functional region. Figure 6 provides comparable
data for Europe in 1996. In the latter we make use of the EU's Nomenclature
of Territorial Units for Statistics (NUTS). The map is based on about 200
regions in what is known as NUTS II.

It should be noted that the circles in all the maps are placed in the geometric centra of the regions in order to indicate that they represent the total number of students in each region, not individual higher educational institutions.[27]

The maps illustrate the dimensions of the Danish and Swedish systems of higher education viewed in an international perspective. They also indicate that higher education is with certain exceptions decentralised. The location pattern follows the distribution of population fairly closely. It should be noticed in passing that the population of the Scandinavian countries is almost equivalent to that of many European and American regions.

When we turn our attention to the regional distribution of economic activities that contain a marked element of research and development, it is evident that there are clear trends towards geographical concentration. Figure 7 provides an example of the high technology industry in Europe in 1995. It has unfortunately not been possible to use NUTS II in this context. NUTS I which has been used here provides us with much more aggregated data where for example Sweden comprises a single region. Figure 8 shows a comparable relative distribution where the number of employees in high technology industries are compared to the total number of economically active persons in the particular region. Thus, a rough but clear outline of the regional strongholds of the knowledge based manufacturing industry in Europe emerges.

When we turn our attention to the regional distribution of economic activities that contain a marked element of research and development, it is evident that there are clear trends towards geographical concentration.

The registration of patents regionally distributed according to the addresses of the applicants has been used in different contexts to indicate where technical renewal actually takes place. Several objections have been raised against the use of this type of data to uncover innovations. Naturally there are inventions that are not recorded in the patent register. Obviously there are also patents that do not lead to commercial products. Moreover there is also a risk that patents will be registered at head offices of major companies rather than at the workplaces where the inventions

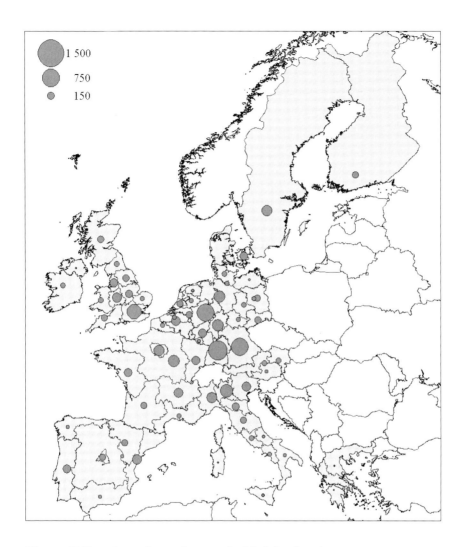

Figure 7 Number of employees in high technology manufacturing industries by regions 1995

were actually made. Nevertheless, American studies for instance indicate that patent statistics are sufficiently reliable to be used in regional surveys. There is also a lack of good alternatives.[28]

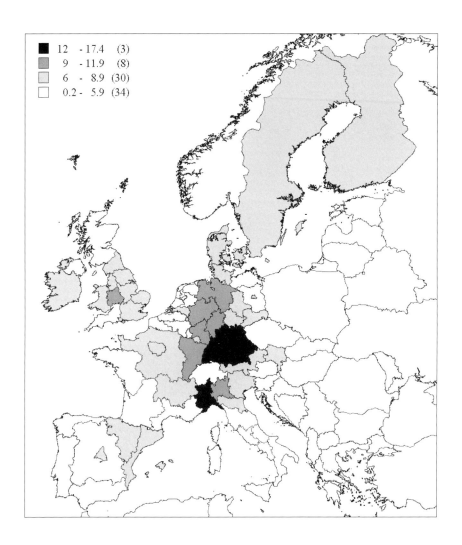

Figure 8 Percentage share of the economically active population in high technology manufacturing by regions 1995

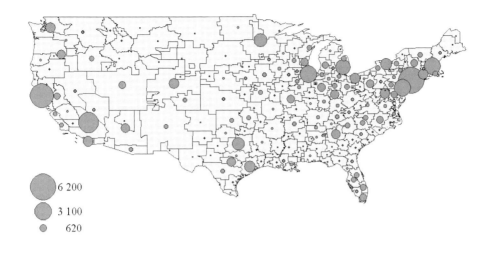

6 200

3 100

620

Figure 9 Number of patents by regions 1996

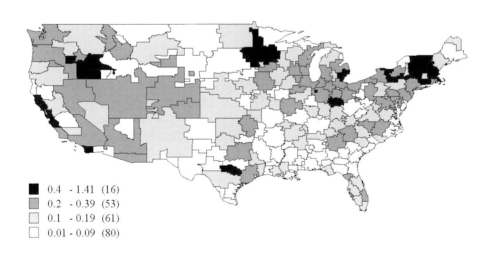

■ 0.4 - 1.41 (16)
▨ 0.2 - 0.39 (53)
▢ 0.1 - 0.19 (61)
□ 0.01 - 0.09 (80)

Figure 10 Patents per thousand inhabitants by regions 1996

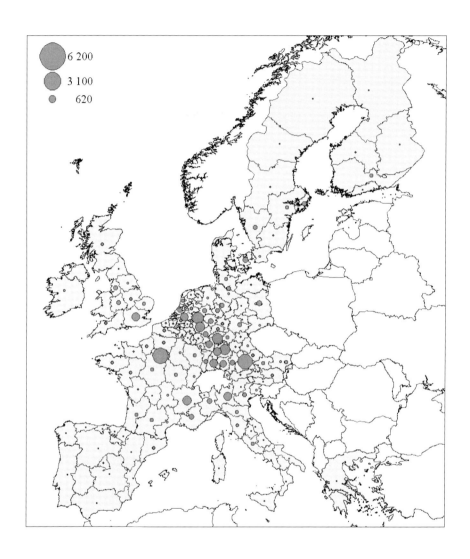

Figure 11 Number of patents by regions 1996

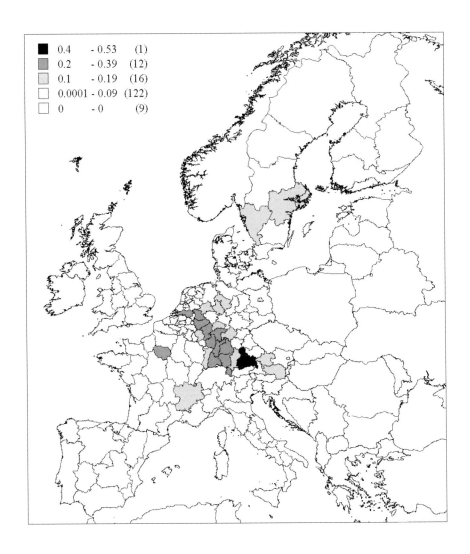

■	0.4 − 0.53	(1)
▨	0.2 − 0.39	(12)
▨	0.1 − 0.19	(16)
□	0.0001 − 0.09	(122)
□	0 − 0	(9)

Figure 12 Patents per thousand inhabitants by regions 1996

According to the picture that patent statistics is able to produce, technological renewal and innovation processes are more geographically clustered than all of the other location patterns illustrated on our maps.

There is an evident cluster along the Atlantic coast in the north-eastern United States and another in a couple of places in California. Around the Great Lakes and in Texas there are also smaller concentrations (Figure 9 and 10). In Europe, the Paris region stands out and there is a marked "innovation ridge" stretching from the Benelux countries to Bavaria. This ridge is equally prominent when the number of patents is compared to the size of the regional population. Paris and Lyons appear as islands. In Sweden two regions are evident in a band that stretches from the West coast to the Lake Mälar valley (Figures 11 and 12). Within these two regions, there is a marked internal concentration to Stockholm and Gothenburg not seen on the maps.

7

A Europe of

Regions

Most people would appear to be familiar with a focus on Europe divided into sovereign states the boundaries of which have established the framework for most political activity. By means of legislation and political action, governments have been able to exert control over economies and citizens have been able to see a relationship between their own welfare and the strength of the country's economy. Within these political frameworks, education and research have also been organised. This is particularly applicable to the Swedish experience. People have lived with national identities that have been formed by upbringing, education and the information that is communicated in unison over wide linguistic areas by means of books, newspapers, radio and television. A uniform geopolitical order has become legitimised and commonly accepted. It has reached into all areas of modern social life, especially during the twentieth century.

In a new millennium, this type of framework is no longer quite as self-evident as it was a few years ago. During the short period that has elapsed since the fall of the Berlin Wall in the autumn of 1989, the role of the territorial state in the future of Europe has begun to be debated and questioned. Researchers and political commentators have pointed to a number of factors which suggest that territorial states have played out the role that they have had for over a century or at least lost some of their traditional hegemony.

Cross-border networks

Networks have become a highly fashionable concept.[29] Manuel Castells who we have referred to previously, uses the concept *Network Society* in order to characterise an emerging world. In this world, many of the most important functions are organised in networks. Networks constitute the new "social morphology" of our societies, and the diffusion of networking logic substantially modifies the operation and outcomes in processes of production, power, culture and human experience. Moreover he maintains that network not only impinge on the hegemony of states but also on other collective spheres of power such as political parties and trade unions. Work is in the process of losing some of its collective identity and by means of extensive specialisation has become more closely associated with the competence and knowledge of individual human beings. A cultural pattern of social communication and social organisation focused on the individual is gaining ground.[30]

What is now happening is that different forms of network are growing in importance at the expense of territories. There is a growing tension and risk for conflict between the power associated with territories and the interests that find expression in networks. The huge growth of institutional networks provide the most marked examples. Company networks and scientific networks that have been of vital importance for economic progress have broken out of their traditional political and social limitations. It is no longer a question of simply adjusting to the size of the market and distant sources of raw materials. A complicated web of cross-border networks are developed as a result of foreign investments, take-overs, mergers, cross-wise ownership, business alliances and co-operative agreements. These arrangements are subject to floating ownership relations and patriotic loyalties. The stateless organisations of the present age can be only partly controlled and influenced within the territorial based decision-making systems. By evolving new strategies for co-operation and specialisation within networks, it is possible for firms to take advantage of differences in the conditions of production in several countries at the same time. The difficulties of conducting economic and labour market policies within national frameworks is one of the most profound challenges to national sovereignty in our time.

What is the scale of these autonomous networks that have broken away from their traditional national frameworks? Data is available for multinational firms i.e. companies and corporations that have operations in several countries at the same time. According to the United Nations World Investment Report from 1993, 37 000 multinational firms and corporations controlled one third of all the privately owned means of production in the world. According to a later report, this figure had increased to 52 000 by 1998. These firms vary in size and type. In addition to manufacturing firms, there are trading and service firms. These data are likely to be an underestimate since it is unable to take account of the vast range of transnational relations between production units that formally belong to different firms and corporations.

It is perfectly understandable that the number and extent of transnational networks is closely associated with the size of the national territory. Large countries that have major domestic markets may encompass wide-scale networks within their boundaries. In small countries on the other hand, there is a tendency for firms to develop networks at an early stage. Let us take the example of a small economically advanced country like Sweden. About 3 500 of the 37 500 companies listed in the UN report in the early 1990s were Swedish. In relation to the size of the country, Swedish firms may be considered unique in terms of their capacity to break out of a national framework.

In the early 1990s, the total global turnover of the twenty largest manufacturing firms in Sweden exceeded SEK 1 000 billion. This figure is equivalent to double the value of Sweden's exports during a year and half of the country's entire GDP. These firms had 775 000 employees which exceeded total industrial employment in Sweden according to official statistics. Somewhere between 69 and 97 per cent of the employees of the major engineering, forestry, pharmaceutical and construction firms which belonged to this group of twenty at that time were located outside Sweden.

The expansion and importance of research is still discussed in terms of national priorities. At the same time, it is undoubtedly the case that research and development is predominantly conducted within international networks outside the control of individual states, despite the fact that many of them still finance a large part of their operations. Science

Science has freed itself from national frameworks and increasingly operates within network structures which are strikingly reminiscent of those that prevailed during the Middle Ages. An archipelago of universities and research institutions are bound together in a network of cross-border relations. The scientific world is now more than ever part of a powerful system of communications.

has freed itself from national frameworks and increasingly operates within network structures which are strikingly reminiscent of those that prevailed during the Middle Ages. An archipelago of universities and research institutions are bound together in a network of cross-border relations. The scientific world is now more than ever part of a powerful system of communications. Research and the development of knowledge is based on the diffusion of ideas and the circulation of information. In a creative process, pieces of information are combined in new frequently surprising ways. Teaching conveys these new ideas to others. Scientific networks and communities are not just used for communicating ideas and viewpoints. They are also used for purposes of control, criticism and recognition.

Renaissance for regions

At the same time as a network society with global dimensions attracts attention, the regional level in Europe has received a growing interest among researchers in the social sciences and humanities. The regional level refers to a level between the national and the particularly local.

In Table 2 we find a rough classification of types of regions together with some examples. In many cases of course, there is an overlap between the different criteria for regional categorisation. Later in this report we are chiefly going to make use of administrative and urban regions in our analysis.[31]

The historical significance of regions may be natural and easy to understand for its descendants. However its assumed role in the development of modern society may be somewhat confusing. An important cause of the regional divisions and isolation of former times was naturally the *friction caused by physical distance*. Water offered the best opportunity for transporting goods and people. By comparison, road transports were limited and of a poor standard. Messengers had to be used to send information. It took weeks and months to send messages over long distances in Europe. In former times, the social communications that were essential for a sense of community and cohesion in a society required close proximity between people. This type of social arrangement prevailed for a long period throughout European history.

Table 2 Types of regions and classification principles

Basis of divission	Principle of division	Examples
Nature	Transportation facilities	Islands, peninsulas, planes, valleys
Culture	Linguistic and ethnic similarity. Shared history and religion	Basque, Catalonia, Wales, Scotland, Wallonie, Sicily, Lombardy
Function	Intensity of flows (goods, people, ideas)	City regions, urban regions, daily urban regions
Administration	Territorial range of decisions and regulations	Bundesländer, départements, cantons, counties

In the shadow of industrialisation and two major wars, political power became centralised in many parts of Europe. The concentration of economic activities affected the settlement pattern. Although the regional and local environment never lost its hold on human senses and continued to provide a framework for everyday life, it was the national centres that established a growing influence over living conditions during most of the past century. As globalisation and a growing economic inter-dependence have become established during the latter part of the twentieth century, a regional level has once again with the help of networks begun to play an important role in the development of European societies at the same time as the national level has lost ground.

There are three different types of regionalisation. One form is equivalent to *decentralisation*. Here power and executive authority have been shifted from the centre of the territorial state to the regional level. This type of decentralisation is fairly widespread in the countries of Europe. The subsidiarity principle of the EU may also be seen as supranational support for this type of development. In the early 1970s, the political system in Italy was subject to major regionalisation. Political power and public service was transferred from central government in Rome to boards and councils in twenty regions. This reform provided researchers with an opportunity to study a rare social experiment.[32] In Spain, Portugal and France, regions have increased their power and influence in relation to central government. The same has occurred regarding the German "Bundesländer". Even in the unitary state that has prevailed in Sweden since the sixteenth century, the regional level has received a newly awakened interest.

Another type of regionalisation is closely associated with cultural conditions and identity that have deep roots in European history. This is a type of regionalism that is obviously opposed to a national central authority; in many cases there are grounds for describing it as *separatism*. In northern Italy, there are forces demanding political independence. The Basque provinces of Northern Spain are an almost classical example. Following a referendum, Scotland and Wales now have their own parliaments. Belgium is on the verge of being divided up. Within both the former Yugoslavia and the former Russian empire, there are striking examples of regional nationalism. The third and final form of regionalisation may be termed *region building* and is best suited to a situation where local and regional forces consciously strive to create a new region or perhaps more accurately try to strengthen an already existing but underdeveloped regionalism in an area.

Decentralising, separatist movements and region building processes may proceed simultaneously and can mobilise mutually supporting forces. This is undoubtedly the case with many of the old border areas that were located out at the periphery of national spheres of interest, far from the country's centre of political power. In the light of the increased centralisation of executive and administrative authority in both the private and public sectors after the Second World War, these peripheral border areas stood out as zones of weakness in the economic landscape. However there are also several cases where peripheral areas have proved to be successful. It is in several of these national border regions that the propensity for change in the Europe of early twenty-first century seems greatest. In a situation where the central authorities of individual states are in the process of losing some of their former hegemony and the European union is expanding, it is in these border regions, far from national centres of power and authority that these changes first become most evident.

The *cross-border regions* that have become established in several parts of Europe during the recent past are especially interesting in this context. Crossing one or several national frontiers, they represent a form of regionalisation that neutralises national boundaries and erodes some of the authority of sovereign states. As a result of earlier conflicts, several of them have been divided up between different territorial states or have changed their national allegiance. There can be residual historical dividing lines that continue to give rise to borderline disputes about borders. However even in the absence of battlefields along national frontiers, there would appear to be a tendency towards cross-border co-operation both within and outside the European Union.

The cross-border regions that have become established in several parts of Europe during the recent past are especially interesting in this context. Crossing one or several national frontiers, they represent a form of regionalisation that neutralises national boundaries and erodes some of the authority of sovereign states.

Many of Europe's former areas of conflict and risk zones have now been converted into areas of co-operation and common development. The scale of this new form of regionalisation has expanded rapidly in recent years. Studies show that there are more than one hundred regional formations of this type in Europe and they would appear to be on the increase. The cross-border regions have formed clusters along almost all of the boundaries of territorial states in Europe. At the same time there are regions that have developed a mutual co-operation without actually bordering on each other.

Along the ancient boundary between Roman and Germanic cultures (*limes*), a long strip of cross-border regions stretches from the Benelux states southwards via the Rhineland, the old conflict zone between French and German interests. Since 1989 in what is rather diffusely called Central Europe or Mitteleuropa, a whole series of co-operative agreements have been established between regions on different sides of the boundaries of territorial states as well as of the former Iron Curtain. Agreements have been signed between German, Czech and Polish regions. Along the Oder and Neisse rivers, Pomerania has been created as a result of co-operation between Poland and Germany. Over a short period, Hungary has developed contacts in the border zones between surrounding countries.

Many of Europe's former areas of conflict and risk zones have now been converted into areas of co-operation and common development.

The Alpine area has also experienced the impact of these types of regional agreements. As a result of Austria's entry into the EU, cross-border co-operation in the eastern Alps has been strengthened and Austria has increasingly become a link between this area and central Europe. The northern part of Italy is clearly linked to its northern neighbours. In the western part of the Alps, there is another zone of old overlapping interests. This is remarkably reminiscent of a geopolitical pattern from the days of the Habsburg dual monarchy. In the Pyrenees on the other hand, the border between France and Spain has been neutralised in places as has also happened further west between Spain and Portugal.

On Europe's northern periphery, a growing Øresund region provides the most interesting example in this context.

In this panorama of cross-border regions, there is naturally a considerable range of regional formations with regard to size and characteristics. What they have in common is that they provide us with an example of regionalisation that does not occur within a national framework. Local and regional actors, both private and public, are usually instrumental in furthering the cause of cross-border regions. The region may be said to be in a grey area between different civil and public laws. Firms, universities, chambers of commerce, trades unions, political parties and cultural institutions are among the most important actors. Co-operation is developed by means of political and administrative networks with local and regional authorities acting as parties. This process is usually supported by some form of umbrella organisation or a network that has a wider geographical scope. Local authorities, regional boards or states are examples of parties that have been included in this process. In the Nordic area, the Nordic Council and the Nordic Council of Ministers have supported efforts to expand the regional dimension. In Europe as a whole, the European Council and the European Union have actively supported the development of regional co-operation.

There are several different motives for the development of cross-border regions and cross-border co-operation. Originally, security concerns were pre-eminent. The early regional ties between Germany on the one hand and Benelux and France on the other have been said to have contributed to peace and stability. Today, these regional formations are seen rather as part of a general European integration process. The support given by the EU to these regional bodies is an indication of their positive role in relation to integration. In addition, regionalisation is said to facilitate economic and physical planning in these numerous border areas. The economic motives are strong. Cross-border co-operation is said to create conditions for growth by enlarging local markets and by a more efficient use of labour and capital. As has been mentioned above, many border areas have experienced a much weaker economic development than in the central core areas of territorial states. Hence this co-operation may therefore be seen as a mobilisation of peripheries.

The regional concept is ambiguous. This is important to emphasise since regions of the same type whether they are identity regions, functional regions or administrative regions will be thereby not comparable. The surface areas, population figures, regional GDP, infrastructure investments and a lot more will vary considerably. The largest functional urban regions in Europe comprise more than ten million inhabitants. In the Nordic area only a couple of regions exceed one million. In Italy, regions such as Piemonte, Veneto, Emilia Romagna, Tuscany, Lazio, Campania and Sicily have populations of about four to five million inhabitants. Lombardy has nine million. Several French regions and a couple of the most successful Spanish regions have similar population figures. Among the German Bundesländer, North-Rhine Westphalia has a population of 17 million inhabitants, while Bavaria and Baden-Würtemberg have populations of 11 and 10 million respectively. In a EU perspective, the Nordic states are "regions" in terms of their population. However in relation to land mass, only France is larger than Sweden. These facts may appear trivial but they are nevertheless important to bear in mind in the light of a continuing discussion on the growing importance of regions in a future Europe.

The New Economic Geography

In contemporary economy, globalisation goes hand in hand with regional revival. A regional recovery is close on the heels of global growth.

"There is indeed a rise of the regional in lockstep with the rise of the global." This relationship has been tested and verified in a number of countries in the OECD.[33]

Agglomerative advantages

It has been well known for a while that forces of agglomeration exert an important influence on the location of different types of production. There are advantages from having production units close to one another, particularly in certain areas and places. The British economist, Alfred Marshall was one of the earliest proponents of the importance of neighbourhood for industrial development. A hundred years ago he observed that in industrialised countries there were areas and places characterised by an industrial atmosphere. "Industry in the air" was his expression. Here similar and related industries congregated forming an archipelago of scattered islands. Marshall put forward three main reasons to explain this phenomenon of industrial concentration and the creation of an industrial atmosphere in a specific place or region.

There are advantages from having production units close to one another, particularly in certain areas and places. The British economist, Alfred Marshall was one of the earliest proponents of the importance of neighbourhood for industrial development. A hundred years ago he observed that in industrialised countries there were areas and places characterised by an industrial atmosphere.

The agglomeration of related firms produce external effects by the creation of a permanent labour market for skilled personnel within a limited geographical area. This area becomes attractive for both employers and skilled labour. The other factor in favour of agglomeration is the growth in the availability of specialised inputs and services. Thirdly agglomeration generates competence.[34]

The Canadian economist, Jane Jacobs, attracted considerable attention in 1984 with the publication of *Cities and The Wealth of Nations*. Here she put forward the view that nations were inappropriate territorial units for an understanding of the ways in which economies operate. Every state is comprised of a mixture of several regional economies. Rich and poor regions are to be found adjacent to each other. Without national policies of regional compensation, the gaps between regions would be unacceptably large. Despite these regional policies, the economic disparities between regions far outstrip those between national average levels. According to Jane Jacobs, the city region is the geographical unit that best provides insights into how economies basically operate. The city regions or urban regions are the proper units in a larger economic landscape. In urban regions, a remarkable amount of economic activity takes place within a small area. Between these agglomerations, the economic landscape is surprisingly empty.[35]

A leading contemporary economist, Paul Krugman, took up Marshall's ideas in his book *Geography and Trade* published in 1991. He argued that states have a role to play in the international economy simply because their governments undertake measures that affect the geographical mobility of goods and factors of production. As a result of political decisions, political boundaries may act as a barrier to trade and factor movements. However there is otherwise no inherent economic sense in drawing a line on the ground and stating that on either side of that line there are two independent economies. For a closer understanding of what is happening in a global economy we must observe what is taking place *within* the boundaries of individual states. If we wish to understand why growth rates differ between countries, we ought to begin by examining the differences in regional economic growth. As has been seen above, external effects and the advantages of agglomeration exert a decisive influence on the localisation of economic activities and the creation of centre-periphery relations. It is hardly likely that the political boundaries will define the space in which these external effects will operate. If we pause, take a step backward and reflect on the most striking feature of the geography of economic activity, we will quite quickly conclude that it is concentration. According to Krugman, the concentration of firms that arise in all industrialised countries may be explained in terms of a Marshallian trinity of factors: a common labour market, access to inputs and services and the transfer of knowledge. All of these factors presumably occur in a city or a small group of towns where people are able to change jobs without having to break up from a familiar environment, where regular face-to-face contacts can be made and where goods and especially services that are difficult to move may be supplied.[36]

In recent years, a terminology has been adopted, particularly in the geographical literature, which confirms the contemporary relevance of Alfred Marshall. Successful regions and localities are said to possess Marshallian benefits. In the most trivial landscape, intelligent regions may break the pattern. In the sea of traditional industrialism, there are "neo-Marshallian islands". To change metaphor, these islands resemble "raisins in a cake". The successful regions and local units that appear in the extensive and varied literature have in many cases widely different characteristics. There are cities, parts of cities, science parks, development centres, technopoles, local production complexes and industrial districts. In literature they exist alongside Italian regions, American states and German länder. What they all have in common is that they form fractions of national territories.[37]

Where are these neo-Marshallian islands or nodes as they are sometimes called? According to Krugman, for example, Silicon Valley is by no means unique in the USA, neither in time nor place. It is simply a somewhat glamorous version of a traditional phenomena. There are in fact a number of important industries in the USA that are heavily concentrated in specific regions or towns. Mature industries that have formed important agglomerations are to be found for instance in Detroit with its concentration of motor vehicle manufacturers, Akron (rubber), Seattle (aviation), Piedmont (textiles), Dalton, Georgia (carpets), Rochester near New York (photographic equipment) and in Hollywood (film). Today attention is focused on the high technology and electronics industries concentrated in Silicon Valley, along Route 128 outside Boston and in the Research Triangle in North Carolina.

In Europe there is hardly a country that doesn't provide examples of regions and localities that possess what we call Marshallian benefits. For instance there are the well known, topical examples of Cambridge-Reading-Bristol, the southern parts of the Paris region, Baden-Württemberg, Grenoble, Toulouse, Montpellier, Sophia-Antipolis near Nice and Santa Croce sull'Arno in Tuscany. Arnold Bagnasco was probably the first person to draw attention to the existence of a "third Italy" in addition to the industrial triangle in the north-west and the neglected Mezzogiorno in the south. This "third Italy" contains small scale, technically advanced and highly productive firms. This region comprises almost all of Emilia Romagna and parts of Veneto. It includes towns such as Bologna, Carpi, Sassuolo and Arezzo.

The embryo for a fortunate concentration of enterprise may in a surprising number of cases be traced to an apparently trivial historical coincidence. Underlying the successes, there are often individuals whose early initiatives started a long process, a spark that ignited a chain reaction. Hence regional success is rarely the result of a conscious, systematic planning.

One of the apparently contradictory features of integrated industrial districts is that they are able to combine competition and co-operation. Firms act as rivals in relation to research and development, renewal and efficiency but frequently co-operate when it comes to administrative services, financing and utilising physical resources. Mutual assistance is normal and technical innovation is diffused rapidly between firms. In a neighbourhood, "external economies" arise which appear to compensate small firms for the absence of scale economies. According to modern theory on firm development, heterogeneity is an important prerequisite for competitiveness. Successful firms have often control over certain resources that other companies lack. These resources are often to be found in the company's own immediate environment. Successful firms are able to do something that their competitors are not able to do as well, as quickly or as cheaply. Development is not about waiting in a world of uniformity and homogeneity. "Little progress would be made in a world of clones".[38]

In this chapter we have hitherto examined the forces of agglomeration that have for long operated in a traditional industrial society. However as has already been stated on several occasions above, it is of the utmost interest for our report that the concentration of activities in a few regions is even more marked in a knowledge-based society. In a newly published report, the American researchers, Edward Leaner and Michael Storper have for example stated that *geographical clustering* is one of the most prominent features in the economic geography of the new millennium. As a result of a highly specialised division of labour in research and development, the need for co-ordination has increased markedly. It is in this context that direct personal contacts play a surprisingly important role in the present era.

"We argue that the Internet will produce more of the same – forces for deagglomeration, but offsetting a possibly stronger tendency towards agglomeration. Increasingly the economy is dependent on the transmission of complex uncodifiable messages, which require understanding and trust that historically have come from face-to-face-contact. This is not likely to be affected by the Internet, which allows long distance "conversations" but no "handshakes".[39]

The driving forces underlying the aforementioned concentration of knowledge and research-intensive production will be discussed at greater length below when we examine the future of the Øresund region.

The regional archipelago

A picture emerges here of a fragmented space, consisting of an archipelago of self-aware regions bound together by different types of network. Several different conditions have interacted to create this new map. The operation of modern transport systems tends to facilitate the development of a nodal settlement system. Strong forces of agglomeration encourage the concentration of productive activities. As we have seen above, firms and institutions within research and cultural life are embedded in regional environments where human beings live and work. However the lack of opportunities to exchange ideas, knowledge and capital over long cross-border distances would impede the growth of entrepreneurship, research and cultural diversity and threaten the individual region with stagnation. Extensive networks become established without an obvious connection to territorial boundaries. The interaction between global forces of change and regional ambitions that exerts such an important influence on our material prosperity is facilitated by the links that bring together a kaleidoscopic world of home bases and places of creativity.

A picture emerges here of a fragmented space, consisting of an archipelago of self-aware regions bound together by different types of network.

"Regions and localities do not disappear but become integrated in international networks that link up their most dynamic sectors."

This is the view put forward by Manuel Castells in his books on the contemporary growth of network societies.[40]

Universities in particular act as strategic links between world-wide networks and local environments. These links communicate in two directions. The university links up a place and a region with centres of knowledge throughout the world. They act as international connection centres. At the same time, the university mobilises local and regional competence in different ways to create an attractive environment in those places where they are located.[41]

Jane Jacobs, argues that the city and its surroundings – the functional city region – is a geographical unit that provides the best insight into how modern economies function. Between the core parts of these regions, the economic landscape is remarkably empty. The French geographer, Jean Labasse presents a similar line of argument in his book *L'Europe des régions.* Economically successful regions are basically nothing other than expansive cities and their surroundings. He also argues that it is the major cities that provide regional identity. He chooses as an example, the north coast of the Mediterranean, which has been frequently described as a dynamic area. However according to Labasse, development in the regions along this coast is entirely concentrated in its major cities. In the Rhone-Alpes region, it is Lyon and Grenoble that are dynamic, in Piedmont it is Turin, in Lombardy, Milan and in Catalonia, it is Barcelona. In Baden-Württemberg which according to many observers is one of Europe's most expansive regions, the growth poles would appear to be the cities such as Stuttgart, Mannheim, Karlsruhe, Freiburg, Heidelberg, Heilberg and Ulm together with some smaller towns.[42]

The city has two characteristics that make it strategic in a geographical perspective. The settlement pattern would appear to be the most stable spatial structure over time in European society. Population, production and all forms of transport are all related to this basic pattern. The long-term view would suggest that it is primarily the cities and towns that have acted as the bearer of European diversity and variation. They represent *geographical continuity* in the midst of the territorial change that we have described above. The other characteristic of the city is that it provides an *interface* between different geographical levels and spheres of interest, where the uppermost level represents the global interests and the lowest level denotes the local.

The most important cities of our time are the product of a combination of technique, method of organising work, the integration of the world of finance and not least the conditions governing social communications. This relationship may also be described as the interaction between networks - physical, institutional, social and cultural. In the city or rather the city region the nodes in these networks are in close proximity to each other. When networks can interact easily with each other, there are good opportunities for synergy effects.

One of the most obvious geographical advantages of the city is that it offers two types of proximity. It offers *territorial proximity* that is equivalent to closeness and neighbourhood. At the same time, it offers *proximity in networks* in relation to other cities. People, establishments and buildings can be brought within easy reach of each other with the help of developed transport and communications systems without at the same time necessarily having to be close to each other. The nodes in the institutional networks attract each other. Administration and the decision-making functions in business, finance, research, interest organisations and public administration are very often concentrated in the major cities. Here clusters of specialised services are developed. News, culture and entertainment are distributed by the mass media from the large urban centres. As a result of the conditions governing social communications, close environments have always been able to offer *meeting places* that are important for creative processes and the diffusion of innovations (see chapter below).

One of the most obvious geographical advantages of the city is that it offers two types of proximity. It offers territorial proximity that is equivalent to closeness and neighbourhood. At the same time, it offers proximity in networks in relation to other cities.

However the city is not just a place for the exchange of goods and services, co-operation and a forum for personal contacts. It is also a melting pot, a place where different life styles and ideas confront each other. It can be at the heart of change. Revolutions usually break out in towns and cities. It is here that new fashions, styles and techniques have first emerged. New types of economic management, methods of organising work and life styles have usually first been developed in an urban environment. New ideas have mostly spread from main cities including those that have originated elsewhere.[43]

THE MECHANISMS
of Regional SUCCESS

We are now confronted with a highly important question for the future of the Øresund region. *What are the factors that generate success in a knowledge-based economy?*

Before continuing with a discussion of this question, it is appropriate to examine the experiences that can be drawn from the literature in this field. At the same time, we can round off by referring back to the discussion that we started in Chapter 4, "Local and Regional Impacts". In addition we are able to gain a more overall perspective of the issues involved.

In our time, economic prosperity, innovative capacity and growth are positive value concepts that are accompanied by characteristics that are highly unevenly distributed throughout the world.

Why are certain areas and places successful in certain respects while other regions and places are not? Why are certain regions attractive to people and productive forces while others are not? Last but not least, what role does research and higher education play in this context?

Several regions that are widely thought of as being dynamic contain successful universities and research institutes. In the international literature, as we shall see below, it is usually large, densely populated regions that are chosen as examples. In Sweden, five regions were highlighted in the mid-1980s as promising areas of growth. These regions, Umeå, Stockholm/Uppsala, Linköping/Norrköping, Gothenburg and Malmö/Lund were characterised by high and stable population growth at the same time as they possessed the country's most significant concentrations of research and higher education.[44] The Swedish debate at both a national and local level has been influenced in recent years by a conviction that there is relationship between higher education and research on the one hand and local and regional growth on the other. This relationship is also considered to apply to sparsely populated regions and smaller towns.

Several regions that are widely thought of as being dynamic contain successful universities and research institutes.

Regional examples

There are several analytical studies of well-known regional environments in the United States, Europe and Japan. The environments have several common features. They are large population centres, contain large universities and research institutes and have a strong element of industrial activities based on high tech, electronics, and information technology. On the basis of these studies, two different groups of environment may be clearly identified.[45]

The first group comprises the *London-Heathrow-Reading corridor*, the *Plateau de Saclay*, south of Paris, *Sophia Antipolis* near Nice, the *Munich region* , the *Kista-Arlanda corridor* and *Tsukuba*, the science town near Tokyo. These areas are characterised by a heavy concentration of high-tech firms and easy accessibility to important universities and research institutes. However the studies that have been carried out have not provided any clear evidence of synergy effects between university research and entrepreneurial success. The contacts between them are few. Despite their close proximity, universities and research institutes live in one world, small, medium sized, and big firms live in another. Finally it should be borne in mind that the regions listed above have undergone rapid growth as a result of comprehensive planning and regulation.

In the other group, we find four different environments that have experienced substantial synergy effects: *Silicon Valley* with *Stanford University*, the *Highway 128 complex around Boston* with *MIT*, *Aerospace Alley* with the *California Institute of Technology* and finally *Cambridge* in Britain with its ancient, prestigious university. Here there are direct links and a substantial transfer of knowledge between university research and clusters of firms. There are numerous institutional and social networks. The networks are frequently held together by key persons who know each other well. The regulatory framework is minimal and the environments are relatively unplanned and have gradually emerged during a long time.

The researchers who have studied these environments suggest that the following factors have played an important role in explaining the remarkable differences between the two groups. It takes ten to fifteen years for synergy effects to appear. Here it is vital that university research is in tune with the needs of industry, as was the case with the space programme and the cold war military-industrial complex. Behind these successes are individuals whose early initiatives began a long run process, a spark that has ignited a chain reaction. Metaphorically speaking, there is a need for a "precision-tooled" interaction between researchers and entrepreneurs. This interaction presupposes mutual understanding and *trust*.

The case studies presented here indicate that there is no simple answer to the question regarding the role played by research and higher education in regional and local environments. The literature in this field provides many good examples of the dynamic role of the university in promoting regional development. However there are also numerous examples of universities and firms that are not dependent on each other at the local and regional level.

If two phenomena occur in the same area – for example successful research and industrial expansion – this could not be seen as providing support for the hypothesis that there is a causal relationship between them. A combination of factors that appear to be successful in one region would not necessarily produce the same effects in another one.

The detailed studies that have been carried out provide fairly clear evidence that the relationships involved are highly complex and that the effects of higher education and research vary markedly between different places and regions. Before proceeding with a discussion of the future of the Øresund region, two interesting observations can be made. Firstly there are the advantages of being part of the mass that is available in large highly dense regions. Secondly, of possibly greater importance, there is the question of co-operation between interest groups and actors that operate in close proximity to each other.

The question of critical mass

In the literature, considerable attention has been devoted to the concept of critical mass. Underlying this concept is the belief that there are economies of scale in research. Usually, the discussion on critical mass is concerned with the actual size of research departments and institutions. It is obvious that research in natural sciences, technology and medicine frequently requires advanced equipment. For economic reasons, this type of expensive equipment cannot be spread around in any way. This is obviously one of the reasons why major American universities and research departments were able to attract highly qualified researchers during and after the Second World War. It is more difficult to say whether or not the same argument regarding critical mass also applies to human resources. The accumulation of competence in one place naturally creates opportunities for innovation. This is borne out by studies of creative environments (see next chapter).

Usually, the discussion on critical mass is concerned with the actual size of research departments and institutions.

However on the basis of the analysis carried out and presented later in this book, it would appear reasonable to assume that critical mass refers to *communication density* rather than the number of persons involved. The number of possible contacts can be expected to vary according to the size of an organisation. However it is difficult to draw any general conclusions in this regard. The greater the number of people that are gathered together at a workplace, the greater the "population density". However this is not the same as an increase in communication density. It is perhaps not always the case that communications operate in an optimal fashion in large institutions and at major universities. A closer examination of the conditions for social communication indicate that they place great demands on co-operation and trust, characteristics that are often easier to create and maintain in smaller groups than in large ones.[46]

However on the basis of the analysis carried out and presented later in this book, it would appear reasonable to assume that critical mass refers to communication density rather than the number of persons involved.

There is also a critical mass at a higher geographical level, at the *societal level* which makes the discussion on critical mass more complex. However this does not make it any the less important. The concept of critical mass does not only refer to the positive internal effects that arise in relation to a particular scale of operations. The mass is also critical in the sense that the scale determines the extent to which external competence an resources are attracted. Hence critical mass is a concept that is closely associated with the new economic geography. As we discussed above in a couple of introductory sections, the globalisation of the economy leads to an increasingly free flow of capital and investment across national borders and between continents. Research as a location factor is not solely a matter of quality but also requires a certain volume in order to be attractive. *Visibility at the international and global levels are dependent on both excellence and size.* Critical mass must accordingly be discussed at several different levels at the same time.

This type of extended geographical perspective on critical mass appeared at several points in the earlier presentation. The question has not been so much one of scale in relation to universities and other scientific institutions themselves but rather the abundance and multiplicity of complementary activities within the local environment in which the university operates. Does the university operate better and are the synergy effects greater in certain environments than in others? Are certain research environments and "scientific regions" more successful than others in being able to attract firms and career-seeking individuals?

In Chapter 6, "The geopolitical perspective", a comparison was made between parts of the United States and the European Union. At the regional level, there are marked tendencies towards geographical concentration in the knowledge-based economy. While higher education is fairly evenly distributed across the population, qualified research, high-technology industries and technical innovation are concentrated in a limited number of places and regions. The maps that accompany this presentation indicate that the more research-intensive a sector is, the greater the degree of agglomeration. An examination of both the volume and quality of research clearly shows that national comparisons miss an important point. *There are actually only a relatively small number of regions and places in Europe and the USA that carry out research at the cutting edge.*

There are studies that show fairly convincingly that the growth of investment in university research does not affect the rate of innovation in industry other than in a few particular places. In sparsely populated regions and in small towns, there is no evidence to indicate that public investments in research affect the rate of technological innovation in industry other than in exceptional cases. In major cities and densely populated regions, there are on the other hand signs that research produces substantial external effects and promotes industrial development.

An examination of both the volume and quality of research clearly shows that national comparisons miss an important point. There are actually only a relatively small number of regions and places in Europe and the USA that carry out research at the cutting edge.

The results of the studies carried out in American city regions are fairly disheartening from a Nordic perspective. The most striking effects of investment in university research are to be found in densely populated regions that have substantial concentrations of highly skilled labour, research-intensive manufacturing industries, new firms, and diversified services. In the USA, such agglomerations have populations of at least one million inhabitants and 30 000 students.[47] In the Nordic counties, there are actually only two city regions that would meet these requirements: Stockholm/Uppsala and a possible future Øresund region characterised by close co-operation between Copenhagen, Malmö and Lund.

Against the background of this foreign experience and in the light of the major studies carried out in Sweden, it is important to point out that it has hitherto not been possible to establish a direct causal relationship between for example the geographical distribution of university research and the location of new innovative industry. The relationships that exist are complex and are "embedded" in that multiplicity and complementarity that is often to be found in and around large urban centres. It is here that we find the *densities* in the various innovation systems and technological clusters, the physical and cultural infrastructure and the highly skilled and narrowly specialised labour.

To extend the discussion, it would be of interest to examine what is actually meant by the concept of critical mass, density and complementarity. It is obvious that several factors *co-operate* to produce significant effects. Each individual factor would not explain very much. The cardinal question is to determine the various components that are absolutely essential in an environment and to ascertain whether compensation can be made for the lack of any particular one.

One important factor has already been discussed. The city operates as a communications centre for all economic and social life. Successful regions and places are products of a combination of the development of knowledge, new technology, innovation systems, integration of financial institutions and not least the conditions governing social communication. These relationships may also be described as a result of the interaction and interplay between networks – physical and institutional as well as social and cultural. In the city or rather city region, the nodes in these networks lie close to each other. When networks may be easily integrated, there is considerable scope for synergy effects.

One of the most obvious geographical advantages of the city is that it offers two different types of proximity. It provides a territorial proximity that gives density and neighbourhood. At the same time, it offers a proximity of networks in relation to other towns and cities.

With the help of modern transport and communications systems, people, industry and buildings can be brought together without actually being physically close to each other. The nodes in an institutional network attract each other. Administrative and executive authority within business, finance, research, interest groups and the civil service are concentrated in major cities. It is here that the clusters of specialised services are developed.

The role of culture in this context cannot be overestimated. Dense environments have always offered meeting places that are vital for social communication, innovation and artistic creation. Cities of varying sizes may be seen as focal points, places where different life styles and ideas are brought into confrontation with one another. These are intense environments where the markets for artistic activities are large and vital.

Theatres, museums and other major cultural complexes have fixed costs that require good attendance. They are often frequently dependent on subsidies and sponsors. For this reason, they are often only to be found in major cities that are able to attract both residents and tourists. Complementarity is also encountered among authors and publishers, artists and gallery owners. Opportunities are created for artists over a wide field to move between different stages, recording studios and occasional venues.[48]

The social web

In 1994, the American researcher, AnnLee Saxenian, referred to above, presented an analysis of IT clusters in Silicon Valley and the Boston region. Her main argument was that it was basically "cultural" differences that explained the greater innovative capacity and growth of Silicon Valley than the high technology centres along Route 128 in the Boston region. The explanation is historical.

The prestigious universities of the Boston region – Harvard, Yale, Princeton and MIT – have for many years been able to take advantage of their well-established relationship with federal government in Washington. A stable contact network that is frequently formal and hierarchical has been developed within the academic world as well as with the public administration and industrial sectors in its vicinity. Developments in Silicon Valley on the other hand have been characterised by a pioneering spirit that to an outsider may appear disorganised, hazardous and unstable. This pioneering spirit in

California has created a regional innovative environment and a development climate at the local level around the university that has stimulated small business enterprises in the IT sector. Risk capital has been available. Co-operation between universities and firms has developed rapidly without any obvious signs of prestige and a formal system of rules and regulations. The attitude towards new businesses associated with the university environment has been encouraging and tolerant. At the same time, it should be pointed out that this environment comprises elements that are conceived as being economically and socially brutal.[49]

The regional examples that have been presented so far are strikingly unanimous in one respect. The most creative economic environments in the knowledge based economy have been developed on the basis of local and regional co-operation between qualified research and enterprising spirit. Individuals with unique skills may combine the roles of researcher and entrepreneur. They are able to move without restrictions between university laboratories research and development departments of the firm. However it is probably more common to establish this close relationship by means of the tightly drawn network of contacts that bind researchers in the academic world with key individuals in the business sector. Local and regional politicians may also play an important role in this context.

In another context, we have referred to the concept of social web, or social fabric in order to indicate a network structure that is local and tightly drawn.[50] We have also put forward the thesis that close, tight networks of this type are probably of strategic importance in a milieu of creativity and places of creation. A short summary of the various perspectives that have been developed may be appropriate in this context. At the same time, it gives us the opportunity to raise the question of whether extensive planning and regulations tend to impede innovative and creative processes.

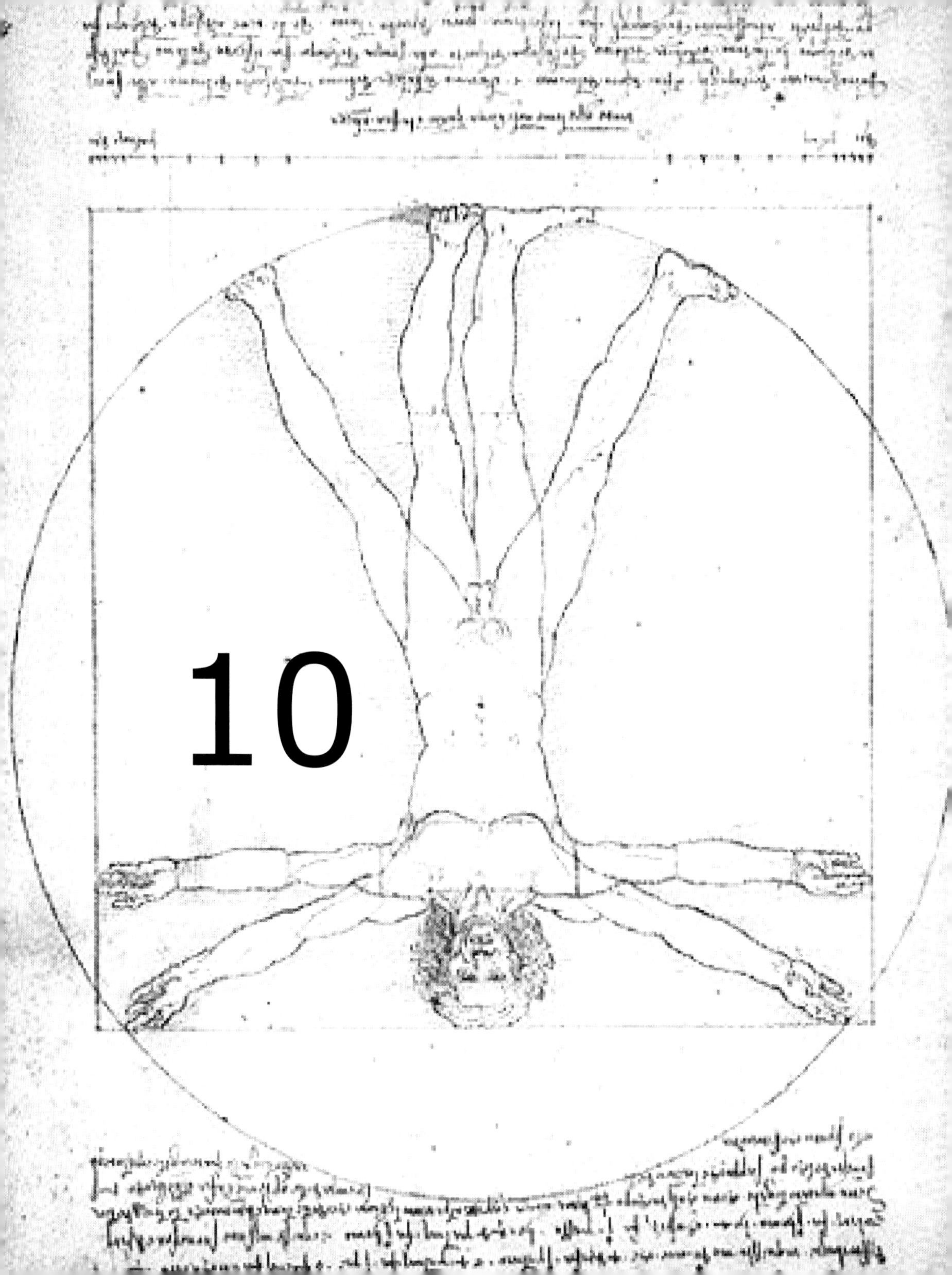

10

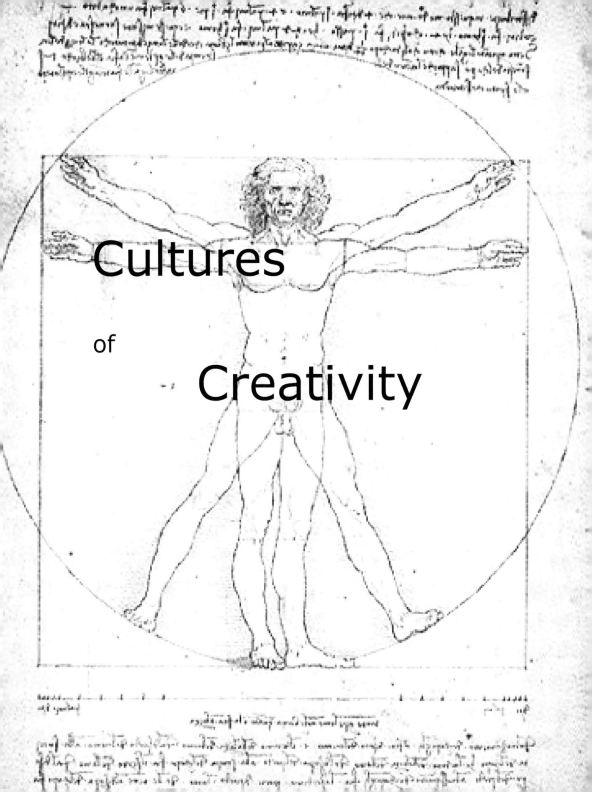

Cultures

of

Creativity

There is an extensive literature on environments or milieux that posterity has considered to be creative. It can be regions, towns or specifically local environments where epoch-making renewal have taken place and inventions been made within for example art, architecture, music, literature, philosophy, medicine science and technology. However it can also be a question of environments in the form of contact networks, organised meetings and institutions within which markedly creative people have worked.

Historical and contemporary examples

Most of the creative milieux described in the literature are historical. It is hardly surprising that it is left to posterity to decide whether or not a certain environment has contributed to a fundamental renewal. In contemporary society, the results of creative processes are difficult to assess, and not to say impossible to record. The historical examples originate from several hundred years before Christ to periods in the second half of the last century. It is worth noting that the older historical examples often span long periods of time, sometimes more than hundred years while examples from our own era are usually much shorter.[51]

Among the examples, we find *Athens* during the fourth century BC, generally thought to be the cradle of Western civilisation. *Florence* provides us with well-known example of Italian urban environment during the late Middle Ages and the Renaissance when it became a centre of renewal in trade, handicrafts, architecture and art. Similar examples could be drawn from the Muslim world of two centuries earlier and from the Netherlands of the sixteenth and seventeenth centuries.

Manchester of the 1840s exemplifies the early years of industrialism. *Vienna*, *Göttingen* and *St. Petersburg* represent some of the cores of radical transformations in the worlds of art, architecture, science, literature and technology during the period around the turn of the century 1800/1900. *The Bloomsbury Group* between 1904 and 1956 allows us to take a long step into our last century. Here we find a part of London that has been looked on as a milieu of creativity in literature, philosophy and economics. However it is not so much the physical environment per se that is considered to act as a stimulus to creativity but rather the close contacts between a limited group of people. The same could be said of Paris and the circle around the bookshop *Shakespeare and Company* during the inter-war years. The depiction of *the Nils Bohr Institute in Copenhagen* is an excellent example of how an institution may operate as a creative environment and how a particular individual may serve as central, cohesive, force in a network of personal contacts. *The Solvay conferences* which at various times have brought together the world's most eminent physicists provide examples of conferences that act as important milieux for scientific renewal. The first was held in 1911 and the twenty-first in 1998.

Apart from the more traditional university environments, research villages, science parks, technopoles and development centres attract considerable attention today. Without becoming involved in a discussion concerning their potential as creative milieux, we may refer again to the *London-Heathrow-Reading corridor*, the *Plateau de Saclay*, south of Paris, *Sophia Antipolis* near Nice, the *Munich region*, the *Kista-Arlanda corridor* and *Tsukuba*, the science town near Tokyo. Among other successful examples, we find *Silicon Valley* with *Stanford University*, the *Highway 128 complex around Boston* with *MIT*, *Aerospace Alley* with the *California Institute of Technology* and *Cambridge* in Britain. The latter along with *Ideon* in Lund which we will discuss later are both closely associated with parts of old prestigious universities.

As was anticipated in the introduction to this chapter, pictures of three different types of environments emerge. There are *geographical* areas and places that are considered to be centres of creativity and innovation. There are *institutions* and organisations within which innovation thrives. In other contexts, it is more appropriate to use the idea of a *network* to describe the way in which contacts are made and innovations diffused. A close analysis however suggests that the differences between the presented environmental forms are illusory.

Geographical milieux entirely predominate the old historical examples. Places and buildings emerge as a physical framework for human creative activity. This seems to be related to the actual durability of the physical environment. Large parts of our knowledge heritage is held in the form of text and pictures and stored in our national archives and libraries as well as in the architecture of our cities. Moreover the geographical environment would appear to have had a greater importance in the past. Physical proximity was a necessary condition for communication. It was within sight and speaking distance that impressions were made and ideas spread. The distribution of information over large distances required lengthy and exhaustive travelling.

Although the geographical milieu may have lost som of its former importance, it is definitely not without influence. As has been pointed out above on several occasions, research has clearly shown that face-to-face meetings and conversations between human beings as well as direct contacts with physical objects continue to act as important conditions for many creative processes.

Firms and universities provide examples of institutional environments, as well as separate places of work and departments. These environments offer us the opportunity to compare differences in creativity between people who work under free and informal forms of organisation and those who are subject to closely defined decision-making hierarchies and rigid regulatory systems. In the literature, this is usually referred to in terms of flat and hierarchical organisational structures.

A network comprising individuals that exert a major influence on each other may naturally have close ties and be difficult to distinguish from the geographical and institutional forms described above. The communications technology revolution has however led to a reduction in the need for physical proximity. Once human beings have learned to become acquainted and develop a community of interests, they are able to co-operate irrespective of the distances involved. Major firms whose plants are widely spread can develop special firm milieux and corporate cultures that are based on a high degree of internal unity. Plants belonging to separate firms and in different branches of industry may develop technological clusters that produce strong internal relationships at the same as the ties to other units are weak. Researchers within the same scientific speciality but located far from each other may have close relationships with one another. Nowadays, power, knowledge and capital are largely communicated within wide as well as compact networks.

Creative processes – irrespective of whether they are technological, research-based or artistic – place great demands on their environment. All environments – irrespective of whether we choose geographical, institutional environments or networks as examples – may act to stimulate or impede creativity. In the following, we will draw attention to the *characteristics* and features that are common to different milieux.

Competence and knowledge traditions

An examination of different environments suggests that there are characteristics in the physical milieu that facilitate creative processes. They may provide meeting places and opportunities for communications. Of possibly greater importance is that they may attract an intellectual elite that places specific demands on its neighbourhood. At the same time, it should be emphasised that it is those individuals who are part of a physical and institutional milieu or belong to a network that actually found the preconditions for a creative process.

In all creative environments, there are human beings that have more or less unique competence. In the literature dealing with such environments there is often an impressive personal index. It is also remarkable that in milieux such as Florence, Vienna, Manchester, London, Paris and St. Petersburg, these skills are simultaneously to be found in so many different specialist fields such as art, music, architecture, literature, science, medicine, technology, philosophy, and political science. It is

> In all creative environments, there are human beings that have more or less unique competence. In the literature dealing with such environments there is often an impressive personal index.

not unusual that these specialists move relatively freely between these different professional areas.

With a few exceptions, skill is based on long tradition of knowledge and extensive experience. The pioneers are usually well aware of the achievements of their predecessors and look at themselves as the latest link in the long chain of knowledge. Different places seem to be associated with different traditions e g music in Vienna, art and architecture in Florence and writing and painting in Paris. A closer examination of these traditions reveals that these skilful personalities often have been imported into these areas rather than having their roots there.

Milieux of creativity ought therefore to be seen as places and institutions that attract human beings who possess unique competence within different areas. The tradition is partly a question of the same places and institutions being attractive permanently during long time.

Among geographical milieux, both historical and contemporary, it is mainly metropolitan centres that predominate. This is perhaps simply due to the fact that it is in major cities artists can find an important market. It is also in metropolitan centres the internal and external possibilities to communicate are the best. There proximity facilitates face-to-face contacts. It is from a central positions in networks that the possibilities to reach out are especially good. Metropolis act as a window on the world. They provide diversity and variation.

Communications

As we have emphasised on a number of occasions above, communication between individuals and areas of competence is of strategic importance in a creative process. In the initial stages of such a process, face-to-face contacts and unexpected meetings that lead to new combinations of pieces of information would appear to be vital ingredients. In both geographical and institutional environments as well as in networks, meeting places

are required for more or less random meetings. In relation to renewal, contacts that lead to new combinations of pieces of information are naturally neither predictable nor able to be planned in advance. Communication using advanced technology will come into its own once a contact relation is established.

Blocking the exchange of information and closing the information channels is a highly effective way of destroying a creative environment. The meeting places that prove to be of the greatest importance are surprisingly often found outside the formal institutions and organisations where the professional and commercial competition is unable to block the free exchange of information.

In ancient, medieval and renaissance environments, the *agora, forum* and *piazza* acted as public meeting places, public rooms. People went on foot along the narrow streets of the city centres. Around the Mediterranean, the climate also offered citizens god opportunity to meet outdoors. Further north, Vienna and Paris had their cafes. Paris also had a bookshop that played a notable role as a meeting place and post office for authors, bohemians and political refugees. In Manchester, the chamber of commerce and various labour movements had particular halls where they held their meetings. Certain bars would appear to have played an important role in Silicon Valley. In other parts of the world, churches, chapels and clubs have also provided meeting points.

The world of science is basically a powerful system of communications. Research and the distribution of knowledge assume that ideas are diffused and that information circulates. As has been noted above, pieces of information will combine in a new, frequently unexpected fashion in a creative process. By means of teaching, these new ideas will be passed on to others. But the scientific community with its networks is not only used as a means of communicating ideas. It also serves as an instrument for control, criticism and recognition. At conferences, work shops and symposia, individual networks are linked up. Bibliometric studies also show how written material disseminates information and facilitates control. The collections of letters written by authors, Nobel prize winners and other cultural figures are an excellent source for analysing communications networks right up to the time when e-mail became the primary source of correspondence. The letters written by Nils Bohr that are registered at the Institute that bears his name in Copenhagen provide us with a good example, demonstrating clearly his central role in the physicist network during the inter-war years.

Diversity, variation and structural instability

Many different factors interact in a creative milieu. In order for synergy effects to arise, some of these factors will have to be available simultaneously in a certain place and be able to interact with each other. It is this demand for interactions at the right time that places such great demands on an environment. As we have seen, this creative environment is frequently culturally diverse, rich in original competence and offers the opportunity for communications both internally and externally.

Diversity and variation would appear to favour creative processes whereas standardisation, uniformity and homogeneity do not. Many of the examples from the literature give an impression that a creative environment is almost chaotic. It is important to note that creative processes and radical renewal are often generated once unique competence and close communication coincide with instability and uncertainty. There is much evidence to support the view that all creative processes – applicable to technological inventions, research at the cutting edge or new art – involves the more or less systematic use of what may be termed "structural instability".

This structural instability makes it easier to break with traditional patterns of thought and rigid regulations. It may well be the case that stable periods of time and planned environments are rarely creative in the word's purest sense. It may be that there is a fundamental contradiction between real creativity and the way we measure efficiency in production or education.

For better or worse, wars and revolutions have throughout history led to numerous innovations and fundamental changes. The economic growth of western economies in the 1950s and 60s was to a large extent based on the creative processes generated during the Second World War. Both the French and Russian revolutions produced new ideas and innovations in areas that had little to do with ideology, politics and different forms of government.

For a period immediately before and after 1900, Vienna was a chaotic milieu where many people suffered. The Habsburg double monarchy was dissolved, the First World War was lost, a civil war was fought and a republic was founded. An almost totalitarian political regime characterised by rigid upper middle class values had prevented social, economic, political and cultural experimentation. A pent-up need to experiment took over sweeping away old authorities and institutions. They were initially replaced by ad-hoc, informal structures. Different styles and ideas flourished at the same time. In the Vienna of that period, one talked of the "Balkanisation" of cultural life.

Manchester of the 1840s also gives a chaotic impression. Both physically and socially, the city was disorganised. What was at that time a major city was largely without stable forms of government. Within education and teaching, there was diversity and variation. The intellectual life of the city was characterised by almost violent, unplanned and disorganised change. It has been pointed out elsewhere that the Manchester of the 1840s is in some respects reminiscent of the Silicon Valley of the 1950s. Similar signs of a lack of institutional forms and social disorder can also be observed in the Paris of the inter-war years, revolutionary St. Petersburg, the Bloomsbury area of London and the nuclear base at Los Alamos in New Mexico.

Even in well planned science parks and technological centres, it has proved difficult to go beyond the preparatory stages of the actual creative process. The components that one believes to be necessary are gathered together in a small area. What happens next is difficult to determine. To an outsider, it appears to be largely random. Occasionally something interesting happens in this contrived environment. However often the synergy effects are largely absent i.e. the results are no more than the sum of the parts. Each separate operation could presumably have generated these results without the major investments in the centralised facilities.

The studies that have been carried out confirm that it is very difficult to deliberately try to construct a creative environment. On the other hand, it is fairly simple to destroy this type of milieu by means of regulations and controls. Looking in the rear mirror, it seems that many of the innovations have been introduced against the intentions of the established institutions and organisations. There would appear to be a marked resistance to the introduction of new ideas in large organisations. Certain studies indicate that successful programmes of change have been implemented once the existing formal organisational structures have been overcome or indeed in some cases actually deceived by the innovators. Here the innovators have gone far outside the formal boundaries of the organisations in order to acquire information, resources and support. They have constructed coalitions and networks outside the formal hierarchies.

11

The Øresund Region

in the European
Urban Landscape

On repeated occasions, we have drawn attention to the substantial geographical concentration of economic activities in the knowledge-based economy. "Geographical clustering" is an appropriate concept to describe this state of things. The economic landscape resembles an archipelago of various city regions where the clusters become more dense. The regions are bound together by physical, institutional, economic and social networks.

Conditions for social communication

An essential driving force underlying the development of location patterns is that the specialists in research and the knowledge-intensive industrial sectors would appear to require face-to-face contacts. This may seem somewhat paradoxical given the communications technologies that characterise our time. However it is presumably due to the fact that the information that is in circulation is not homogeneous.

Recent studies that have emphasised this paradox point to the necessity of distinguishing between two different types of information. *Codifiable messages* are able to be expressed in the form of text e g written instructions and manuals. This type of information is easily disseminated by the use of modern communications technology. *Uncodifiable messages* are dependent on the meetings that take place between human beings, continuously and preferably over a long period of time. This type of information is best communicated between individuals who feel trust and understanding for each other.[52]

Modern computer-based information systems enable us to process an almost infinite amount of information. Using telecommunications technology, information may be transferred at the speed of light over large distances. At the same time, as technological developments have opened up new opportunities for transferring information, the need for consultation and direct personal contacts has become greater. A growing number of private and public sector employees devote an increasing amount of time to meetings and other types of contact activities. Contact trips, even over long distances have markedly increased during recent years, despite the substantial costs in terms of money, time and effort that are associated with travel. This type of contact activity is not just concentrated to the major cities, it has also forced the extensive development of rapid passenger transport facilities.

Well structured, routine information may be transferred rapidly and efficiently by means of telecommunications equipment such as the telephone, fax and computers in networks. This type of information transfer is now a requirement for management, regulation and control in complex transport and communication systems. This type of information is used in the ordering of tickets, flight check-ins and in the banking system where business may be carried out instantaneously between computers which may be in the same building but could just as well be anywhere in the world. This type of information controls manufacturing processes and allows firms and public authorities to administer units that are widely geographically scattered. The information flows are often one way and follow well-established formal channels. Uncertainty is limited (Figure 13).

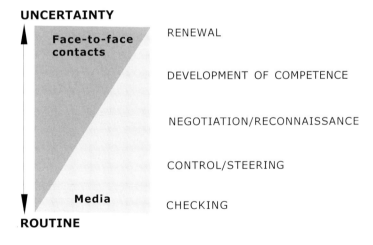

UNCERTAINTY

Face-to-face contacts

RENEWAL

DEVELOPMENT OF COMPETENCE

NEGOTIATION/RECONNAISSANCE

CONTROL/STEERING

Media

CHECKING

ROUTINE

Figure 13 The aims, characteristics and form of information transfer

Technological equipment will be inadequate in the face of the need to tackle questions concerning uncertainty, unpredictability and surprise. There is information associated with negotiations, orientation and search. There is information that form basic elements in processes that lead to the formation of knowledge and innovation. Major research has shown that the exchange and circulation of these types of information mainly occur as a result of face-to-face contacts and group conversations. Research demonstrates that communications technology may be used to transfer information within already existing social networks where uncertainty is limited. However they cannot replace direct personal contacts between persons who are not known to each other and between networks where uncertainty is considerable. Above all we can see the importance of direct personal contacts in creative processes – irrespective of whether they refer to technological development, research or different forms of artistic activities. The question of differences in preconditions and barriers in relation to these processes in various environments has already been dealt with in Chapter 10 entitled "Cultures of Creativity ".

However it is not only the actual characteristics of different forms of communication that explains this development.

The far-reaching division of labour and specialisation has meant that society has become increasingly complicated and difficult to oversee. In a society where knowledge has become fragmented and where a large number of specialists are sitting with a small piece of information, meetings become a means of drawing together diverse information in order to gain a holistic perspective and a basis for decision-making in an uncertain world. It is in the light of these experiences that an examination of the availability and accessibility of the Øresund region should be seen.[53]

The Øresund region as a transport node

The opportunity to travel in Europe has changed dramatically during recent decades. It is not possible to examine all of the different aspects of the various changes that have taken place in the European transport network. Instead the discussion here will concentrate on the travel opportunities between 97 city regions in Europe, all of which have at least 500 000 inhabitants with the exception of certain capital city regions. The regions are functional, consisting of a central city and in certain cases conurbations together with their hinterland. The regions may be termed labour market regions where the majority of the population commutes between home, work and daily services without crossing the region's borders. These regions are referred to as Daily Urban Regions. The circle symbols on the maps in Figures 14 and 15 indicate the major city in each region.

The size of the symbols depict the inbound and outbound accessibility of different centres in the European urban system. *Inbound accessibility* provides for each travel goal the answer to the following question "What proportion of the inhabitants in the 97 city regions can reach this point for a day return visit?" The rule here is that the journey must be taken between 06 00 and 24 00 hours and allow at least 4 hours stay at the destination. It is assumed that the most rapid form of transport or some combination of boat, car, train, flight is used. Boat is only used in a few cases such as between Malmö and Copenhagen prior to the construction of the bridge/tunnel. Cars are used for distances up to about 150 km whereas trains are more frequently used over distances up to 600 km. Flights tend to be chosen for distances exceeding 600 km. These estimates are repeated for all of the 97 cities. The estimates of *outbound accessibility* assume that each city acts as a starting point. The question is then "What proportion of the population in all regions can be reached from this point?" The same conditions apply to these estimates as above.

Two conditions exert a major influence on the results: the region's own population and the proximity of major concentrations of population. Good direct rail and especially air communications provide a substantial surplus, even over long distances. There are many different ways of carrying out the estimates. It is then important to note that the results of the different estimates often agree with each other. Here we present two representative maps of the inbound and outbound accessibility of different centres in the European urban system in the early 1990s.[54]

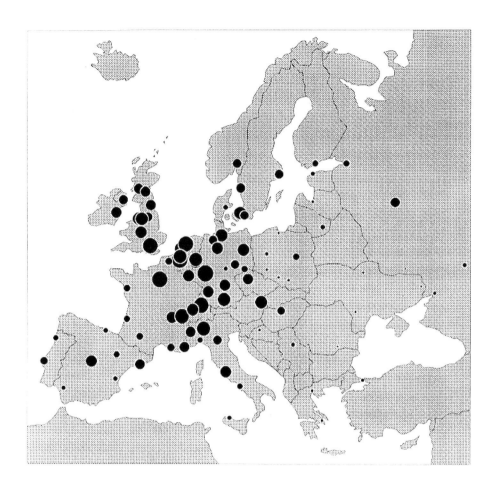

Figure 14 Inbound accessibilities in the European urban system

As the two maps indicate, the inbound and outbound accessibilities of the European urban system vary markedly between city regions. The most advantageous positions are to be found in an arch that stretches from the British Isles southwards via the Rhine valley and down to northern Italy. This arch has been called Europe's economic backbone,

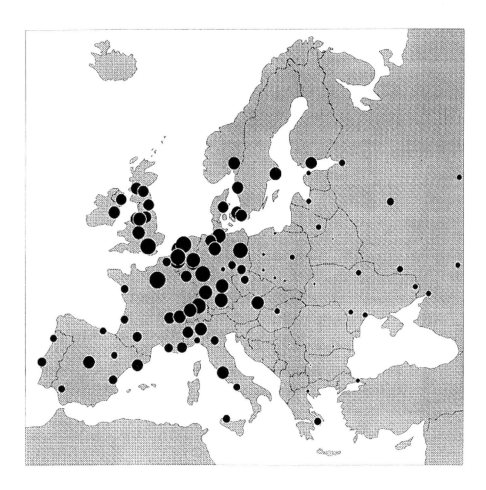

Figure 15 Outbound accessibilities in the European urban system

the Blue banana, the Boomerang and the Baroque arch. Alongside this arch, Paris and several regions in Germany offer excellent travel opportunities. The 15 best situated regions are presented below in Table 3. Additional data is also available for a few less central regions that are nevertheless of interest from a European perspective.

Table 3 Centralities in Europe 1992. Inbound and outbound accessibilities (index)

Urban centres	Inbound accessibilities	Outbound accessibilities
Frankfurt a. M	100	92
Paris	94	100
Brussels	92	93
London	85	93
Amsterdam	84	96
Zürich	84	83
Genèva	74	61
Milano	71	60
Düsseldorf	67	77
Rotterdam	64	85
München	63	73
Manchester	60	62
Hamburg	58	73
Wien	58	61
Copenhagen	58	58
–		
–		
Gothenburg	34	53
Oslo	31	57
Stockholm	30	58
Malmö	29	51
–		
–		
St Petersburg	17	18
Helsinki	16	53
Aarhus	11	49
–		

Source [55]

Outside these central parts of Europe, there are large areas, especially in the east that from a communications standpoint may be seen as peripheral. As Table 3 clearly indicates, Copenhagen is fairly well situated in the European city landscape. It ought to be noted that the estimates presented here were made before the fixed link between Copenhagen and Malmö was completed. This link

As Table 3 clearly indicates, Copenhagen is fairly well situated in the European city landscape.

should have further improved Copenhagen's situation. In other parts of Scandinavia, good air communications provide several cities with semi-central positions despite the great distances involved.

As may be seen from the maps and table, a number of cities show clear differences between inbound and outbound accessibility. The most important explanations are to be found in the air timetables and the distribution of population. Roughly speaking, the contact possibilities from the peripheral areas to the central city regions are better than the opportunities in the opposite direction. Stockholm and Helsinki may serve as examples.

Industry
and
Research

12

in the Øresund Region

In an introductory chapter to this book, a brief presentation was made of the Øresund region (see Figures 1 and 2). It was stated that the area comprised a population of about three million inhabitants and half as many workplaces. It is therefore the largest and most densely populated region in Scandinavia; the Stockholm region is second. Four aspects of this cross-border region are of particular interest. Firstly there is the question of accessibility discussed above. Secondly there is the marked concentration of industry on either side of Øresund, especially in research-intensive sectors. Thirdly the region is particularly committed to research and higher education. Finally the region has seen a notable development of networks and collaborative arrangements in recent years.

Industry

The population censuses of 1960 and 1970 indicated that the working population on the Danish side of Øresund, i e Zealand, was two and a half times larger than the working population on the Swedish side, i e Scania. This was particularly marked in relation to typical urban services such as bank and insurance, public administration and public services, transport, trade, hotels and restaurants. The difference between the manufacturing sectors on both sides of Øresund was not as marked as we have seen for services. Employment in agriculture and forestry was more or less the same on both sides of the national boundary.[56]

During the first half of the twentieth century, manufacturing employment expanded in traditional sectors on both sides of Øresund such as textiles, clothing, shipbuilding and food-processing related to agriculture. During the 1950s and 60s, these sectors went into drastic decline to be replaced by the growing urban-based services. In addition, the composition of manufacturing industry has undergone changes.

The geographic distribution of manufacturing in the Øresund region in 1997 is shown in Figure 15. The data behind the map is based on official

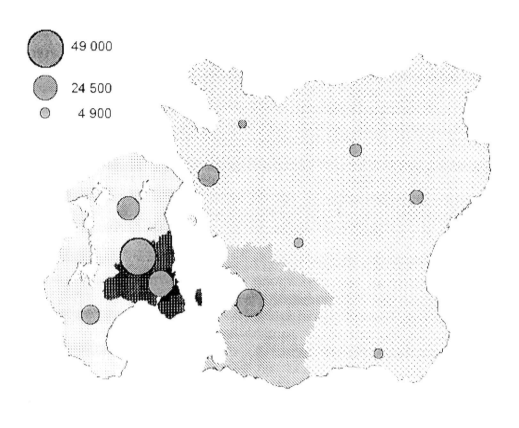

Figure 15 Employment in manufacturing industry in the Øresund region 1997

Swedish and Danish industrial statistics. On the Danish side, the data covers central Copenhagen, the county of Copenhagen and outer Copenhagen which is equivalent to the rest of Zealand. On the Swedish side, the data distinguishes between two areas: the Malmö-Lund labour market region and the rest of Scania.

In order to carry out a more detailed analysis of the composition of industry in the area, a more detailed sector classification has been used based on primary official data sources.[57] This classification corresponds to the questions that have been raised in this report.

The following presentation is based on a report written by Karl-Johan Lundqvist and Lars Winter at the Universities of Lund and Copenhagen where an ambitious attempt has been made to establish a comparable platform for the Danish and Swedish data. The five principal sectors which have been created according to this classification and are considered to be comparable are shown in Table 4.[58]

The research-intensive industry (R&D-sector) is characterised by production that has sizeable elements of research and development, measured both in terms of share of value added and the share of employment of personnel with formal Ph D qualifications. The sector has almost double the engineer density as the knowledge intensive sector which on its part employs substantially more highly skilled personnel than the capital intensive, labour intensive and sheltered sectors.

Previous studies have shown that it is the research-intensive sector that is responsible for most of the innovations in Swedish manufacturing industry. It is here that the new employment opportunities are created and new products developed for foreign markets. It is the most global of all of the industrial sectors and has market segments that are far from their home base in Sweden. This sector is obviously of major importance in relation to an assessment of the future of the Øresund region.

Table 4 Industrial sectors

R&D-SECTOR   RESEARCH AND DEVELOPMENT INTENSIVE SECTOR

This sector is characterised by major investments in research and development. Heavy demands are placed on advanced production factors and external economies of scale. This sector has also heavy demands on local markets, the availability of highly educated labour and advanced services. Many products are at the start of their product cycle.

K-SECTOR   KNOWLEDGE-INTENSIVE SECTOR

Characterised by high technology/engineering intensity. Heavy requirements in relation to external scale economies and supply of engineers. In other respects, low demands on production factors. Internal scale economies. Several products in the later or mature phases of product cycle.

C-SECTOR   CAPITAL INTENSIVE SECTOR

This sector is characterised by extensive raw material inputs and a high capital intensity. Limited demands on labour and other production factors. Most products are in the later or mature phases of product cycle.

L-SECTOR   LABOUR-INTENSIVE SECTOR

High employment in relation to value added. Dependent on cheap labour. Fewer demands on other production factors. Most products are in the later or mature phases of product cycle.

S-SECTOR   SHELTERED SECTOR

Requirements and characteristics as in the labour-intensive sector. Output was previously largely directed towards limited national and regional markets. Indirect protection from transport difficulties, regulatory system and legislation rather than by means of formal barriers to competition such as tariffs, subsidies and technical barriers to trade. This sector is now becoming increasingly subject to competition.

Table 5 Percentage share of industrial employment by sector 1997

	R&D-sector	K-sector	C-sector	A-sector	S-sector	Total
Sweden	16	30	13	22	19	100
Denmark	12	25	72	62	9	100
Stockholm region	44	20	3	9	24	100
Øresund region	19	20	10	21	29	100
Malmö-Lund	18	26	5	18	33	100
Rest of Scania	7	21	16	29	27	100
Copenhagen region	24	18	9	19	29	100

Source [59]

Table 5 shows the composition of industry in the entire Øresund region as well as in its Swedish and Danish components. In order to assess the relative position of the region, data is provided for Sweden and Denmark as a whole. A comparison with the Stockholm region is also considered to be of interest in this context.

The centre of gravity of Danish manufacturing industry lies in the labour-intensive and sheltered sectors. The major food processing industry is to be found in the latter sector although from a Danish perspective, this industry is not sheltered but subject to substantial competition. The knowledge-based sector predominates in Swedish manufacturing industry. It is here that we find a number of Sweden's major engineering firms. The forest products industry in particular contributes to the greater role of the capital intensive sector in Sweden. Finally it may be noted that the research-intensive sector has a somewhat stronger position in Sweden than in Denmark.

The concentration of the research-intensive sector to the Stockholm region is remarkable. Up until 1997, no other labour-market region in

Sweden has succeeded in attracting research-intensive industry in a comparable fashion.[60] Neither Copenhagen nor the combined Øresund region has been able to show anything along similar lines.

In addition to this regional cross-section analysis, it is also important to examine the changes over time (Table 6). Despite the shortness of the time period, certain clear features emerge. Industrial employment in Denmark has contracted between 1992 and 1997. In Sweden, there has been a slight increase over the same period. This reflects a certain recovery in industrial employment in Sweden compared to the previously observed decline between 1990 and 1996 (see Table 1).

This period in Sweden is particularly marked by a shift in employment towards the research-intensive sector. This shift becomes even more evident when viewed in terms of value added and exports. In Denmark as well, it is the research-intensive sector that records the highest growth rates in this period.

Table 6 Percentage changes in employment 1992 - 1997

	R&D-sector	K-sector	C-sector	A-sector	S-sector	Total
Sweden	19.5	9.4	- 2.2	13.7	- 3.6	7.3
Denmark	11.7	3.6	- 6.2	- 5.4	- 4.8	- 1.2
Stockholm region	11.2	21.7	23.1	48.5	4.5	14.3
Øresund region	9.5	- 0.4	- 3.2	- 2.2	- 80	- 1.7
Malmö-Lund	52.2	24.9	- 2.9	13.6	- 6.2	12.5
Rest of Scania	35.8	16.1	17.6	2.0	- 6.6	6.2
Copenhagen region	1.7	- 12.8	- 13.8	- 7.9	- 9.0	- 7.6

Source [61]

The regional shifts are worthy of particular attention. It should be noted en passant that the relative growth of the C- and L- sectors in the Stockholm region ought to be seen against the low absolute values for these sectors at the start of the period. From a position where these sectors were almost entirely eliminated in the capital region in the early 1990s, a slight recovery has taken place in absolute terms.

The authors of the report referred to here view the high relative growth of the research-intensive sector in Malmö-Lund (52 per cent) and other parts of Scania (36 per cent) as a result of a hierarchical diffusion of growth in the Swedish regional system. Stockholm which has been the engine of recent developments has started to experience a degree of saturation in the form of for example, a housing shortage, high costs and crowding effects which have tended to reduce and divert growth to other areas. From a Danish perspective, regional development has been different from that in Sweden. Aggregate manufacturing employment has declined in Denmark. This decline is particularly evident in the Copenhagen area. Growth in the research-intensive sector has been slower in Denmark than in Sweden and has occurred in Jutland rather than in the Copenhagen area, somewhat surprisingly compared with the development trends in Sweden.

Leaving the manufacturing sector, there is unfortunately a lack of comparable data in the service sector. Here there is a great need for further research. It appears likely that there has been a substantially greater growth of service employment on the Danish than on the Swedish side of Øresund. This is confirmed by the data presented in Table 7 which relates to the growth of service employment in the IT sector. The latter is especially important since it provides the advanced knowledge-based services that are demanded by both the research sector per se and those industrial sectors that have become leading innovators.

As the table indicates, the proportion of IT service employment is fairly low in relation to the total volyme of employment. Its most prominent characteristic is its major geographic concentration. Of total IT service employment in Sweden and Denmark 39 per cent was located in Stockholm and a further 28 per cent in the Øresund region. From the perspective of this study, it is important to note that the Danish side of Øresund is completely predominant in terms of the location of IT service employment. *Copenhagen is a major service centre.* Further studies will be able to ascertain whether or not this applies in a more general sense which the 1960 and 1970 census data lead us to believe.

Research and higher education

At the beginning of the year 2000, there were more than 130 000 enrolled university students in the Øresund region together with a further 10 000 researchers at the departments of the universities in the region. In a European perspective, this is a fairly modest figure. However, this concentration of research and higher education is the highest in the Nordic countries. By comparison, Stockholm and Uppsala have only half as many students and researchers.[63]

Table 7 Employment in the IT service sector 1997

	Number of employees	IT sector's % share of total employment in each area	Area's % share of total IT empoyment in Sweden and Denmark
Sweden and Denmark	75 521	1.2	100.0
Sweden	48 390	1.3	64.1
Denmark	27 131	1.0	35.9
Stockholm region	29 550	3.6	39.1
Øresund region	20 962	1.5	27.8
Malmö-Lund	2 474	1.1	3.3
Rest of Scania	565	0.2	0.7
Copenhagen region	17 922	1.9	23.7

Source [62]

In order to compare the volume of research on an international basis, we have chosen to use the available measurements of scientific output. Given the nature of this report, data on students and researchers has not been considered to be of the same interest. In the first step that is presented here, we have chosen to present a comprehensive analysis of major research areas of special interest for this report, without taking account of specific disciplines.

The number of publications in approved academic journals and conference proceedings has been a widely used measure of performance in recent years. In *bibliometrical studies*, an estimate is made of the number of published articles in leading academic publications in different areas. The origin of the articles is recorded with the aid of the published addresses of the university departments where the authors work. This type of approach may be criticised on several grounds. At the same time, it is the only available means of carrying out these extensive quantitative comparisons.

Measures are available for the number of published research reports in relation to GDP for a range of countries. Here Sweden is in second place after Israel and on a level that is twice that of the USA and far in excess of the average for the industrialised countries. The academic journals studied here belong to the fields of science, engineering and medicine.[64] However these nationally based comparisons miss an important point. There are actually only a very small number of regions in different countries that provide really high quality research.

It is geographically concentrated in a manner that is more or less reminiscent of the distribution of patents that was discussed previously in this report (Figures 9 and 11).

The data in Table 8 has been taken from a recently published study carried out by a couple of Danish researchers Christian Wichmann Matthiesen and Annette Winkel Schwarz. Here the bibliometric statistics have been collated on a regional basis. The table covers 39 urban regions (city regions) in Europe all of which have nationally important univer-

Table 8 Number of published academic articles 1994-96, by urban regions

Region	No of articles	Region	Per 1000 inhabitans
London	64 742	Cambridge	81
Paris	45 752	Oxford-Reading	41
Moscow	39 903	Geneva-Lausanne	29
Amsterdam-Haag-Rotterdam-Utrecht	36 158	Basel-Mülhausen-Freiburg	20
Copenhagen-Lund	21 631	Bristol-Cardiff	15
Stockholm-Uppsala	20 195	Zurich	13
Berlin	19 872	Stockholm-Uppsala	12
Oxford-Reading	18 876	Helsinki	12
Edinburgh-Glasgow	18 688	Copenhagen-Lund	11
Manchester-Liverpool	18 653	Amsterdam-Haag-Rotterdam-Utrecht	10
Cambridge	17 764	Munich	10
Madrid	16 230	Edinburgh-Glasgow	10
Munich	15 947	Gothenburg	10
Dortmund-Düsseldorf-Cologne	15 716	Mannheim-Heidelberg	8
Milan	15 120	Oslo	8
Rome	15 088	London	7
Frankfurt-Mainz	14 512	Lyon	7
Basel-Mülhausen-Freiburg	13 918	Milan	6
Sheffield-Leeds	13 484	Frankfurt-Mainz	6
Geneva-Lausanne	13 405	Prague	6
Mannheim-Heidelberg	12 289	Dublin	6
Zurich	11 951	Paris	5
Brussels-Antwerp	11 786	Berlin	5
St. Petersburg	11 506	Rome	5
Barcelona	11 467	Brussels-Antwerp	5
Vienna	10 882	Sheffield-Leeds	5
Bristol-Cardiff	10 633	Vienna	5
Helsinki	10 287	Manchester-Liverpool	5
Birmingham	9 882	Barcelona	5
Aachen-Maastricht-Liège	9 705	Aachen-Maastricht-Liège	5
Lyon	9 175	Birmingham	5
Warsaw	7 966	Madrid	4
Prague	7 616	Warsaw	4
Hamburg	7 425	Stuttgart	4
Gothenburg	7 378	Moscow	3
Budapest	6 697	St. Petersburg	3
Oslo	6 466	Hamburg	3
Stuttgart	5 043	Budapest	3
Dublin	5 043	Dortmund-Düsseldorf-Cologne	1

Source: The Sience Citation Index [65]

sities and research institutes. The source for the Danish study is a data base containing data from 5 000 leading academic journals in science, engineering and medicine, The Science Citation Index (SCI). It has been developed by the Institute for Scientific Information (ISI) in Philidelphia in the United States.

Measured in *absolute* figures, London, Paris, Moscow and the Amsterdam–Haag–Rotterdam–Utrecht conurbation are the most important research regions in Europe. These are followed by Copenhagen –Malmö–Lund and Stockholm–Uppsala which is fairly remarkable viewed against the background of the geographic distribution of students enrolled at European universities (Figure 6).

In *relative* terms, taking into account that urban regions vary markedly in population size, Cambridge emerges as the most successful research centre in Europe with about double the amount of academic production per capita compared to its neighbour Oxford–Reading. The position of Geneva–Lausanne is tied up with activities of the CERN research laboratory. Stockholm–Uppsala, Helsinki and Copenhagen–Malmö–Lund are highly placed while large densely populated regions such as London, Paris and Moscow are far down the list.

Networks and forms of collaboration

The former vice-chancellor of Lund University, Håkan Westling, describes in his book *Idén om Ideon – en forskningsby blir till* ("The Idea of Ideon – the Emergence of a Research Village") the foundation and development of a science park on the campus of Lund University and the Lund Institute of Technology. It was founded in 1968 and is now a very good example of what can be created by close *collaboration* between the university and the business sector.

During the life of the science park, 435 firms have been involved. 166 firms were still in existence in the year 2000. Most of those firms that have left Ideon have quite literally outgrown an environment that was meant to operate as an incubator for newly born firms. Several have been restructured and changed name. Others have ceased to operate since the original commercial idea was no longer profitable. Approximately 30 firms have gone bankrupt. Eleven have been floated on the stock exchange. Others are preparing an introduction. It is perhaps relevant in this context to remind our readers of the concept of structural instability that was discussed in the section dealing with "Cultures of Creativity".

Of the firms 38 per cent are engaged in information technology and a further 35 per cent in the biotech and pharmaceutical fields. Consultant and service companies comprise 16 per cent. Most firms are small, are subsidiary firms or form the research departments of major companies. IT firms are generally much larger than the others which accounts for their high employment share within Idéon (63 per cent). The telecommunications giant, Ericsson and the pharmaceutical company group AstraZeneca are among the well-known enterprises that have major plants outside but close to the science park.[66]

In an earlier chapter, it was mentioned that the initiative for collaboration in cross-border regions was frequently taken by local and regional actors, public and private. The formation of such a region may be said to occur in a grey zone between civil and public legal jurisdiction. Firms, universities, chambers of commerce, trades unions, political parties and cultural institutions are among the actors. Collaboration is developed by means of political and administrative networks within which local and regional authorities operate. It is often quite normal that this process is supported by some form of umbrella organisation or a network that has a greater geographical scope. In this way, complex network structures develop that bind together operations on both sides of the boundary between sovereign states.

Informal networks have been established by means of personal contacts and usually function alongside the formal decision paths and institutions. At the same time as outsiders find them difficult to identify, they can lead to a dramatic reorganisation of traditional and accepted power structures. *Formal networks* on the other hand do not present a challenge in the same way to old established power structures. They are often created by agreements that are soundly based in established institutions. These agreements and the composition of networks are often well documented. In an integration process of the type that can be studied in the Øresund area, both formal and informal networks operate parallel to each other. They are presumably so intertwined that it is impossible to separate them. We have decided to have a closer look at a number of networks that are formal in character but also have interesting informal features.

THE ØRESUND COMMITTEE is comprised of politicians and administrative personnel. On both the Swedish and Danish sides, they represent municipal authorities, communes and county councils. The present chairman comes from the regional administrative board in Scania. From January 1st 1998, the regional association for Scania ("Region Skåne") has taken the place of the county councils on the Swedish side. The Swedish and Danish governments participate through observers. The committee receives economic support through the EU. Its function is to stimulate cross-border co-operation and has a major role to play in relation to research, culture, education, the environment and the labour market. The Øresund Committee is a good example of a network that has strong territorial ties to the political administrative systems at local, regional and national levels.[26]

THE ØRESUND UNIVERSITY is on the other hand an example of what we previously called an institutional network that is to an increasing extent autonomous of territorial power structures. It can be described as an association or federation of twelve universities and university colleges in the Øresund area. Together they employ 10 000 teachers and researchers and have 130 000 students. There are many nodes in the network; university departments, research institutes, administrative units, laboratories, data bases and libraries are all brought together into a computer-based network. Students are able to follow combinations of courses provided by university departments on both sides of Øresund. Cross-border research co-operation takes place in informal networks. At present there are a number of joint projects under way that will examine the emergence of the Øresund region as a major social experiment. In order to co-ordinate these activities, there is a formal board which consists of the vice-chancellors and rectors of the various units that are part of the federation, an executive committee and a joint secretariat.

MEDICON VALLEY ACADEMY is a network organisation, a cluster, that involves both firms and universities in the Øresund area. This formation is of particular relevance to this book and we can here refer back to the discussion that we had in relation to Figure 4 (p 41). The purpose of the organisation is to mobilise and utilise the material and human resources that were previously divided by a boundary. It encompasses in addition to many of the university departments, 26 hospitals, 100 bio-tech firms, 125 medical technology companies and 71 pharmaceutical firms as well as a series of financiers who have invested risk capital.

The major foodstuffs industry complex around Øresund comprises agriculture, food manufacturers and packaging industries. Together with research departments, this complex forms one of the world's largest concentrations of competences in the field of food technology. It is bound together in a network organisation, the **ØRESUND FOOD NETWORK.** Environmental research is another area that has a substantial concentration of operations on both sides of the national boundary under the heading **ØRESUND ENVIRONMENT**. The predominance of IT in the Ideon science park has already been discussed. Account has also to be taken of the IT industry and research on the Danish side of the Øresund. All of these fields are brought together in a cross-border network, **IT ØRESUND**, that coordinates their diverse activities.[67]

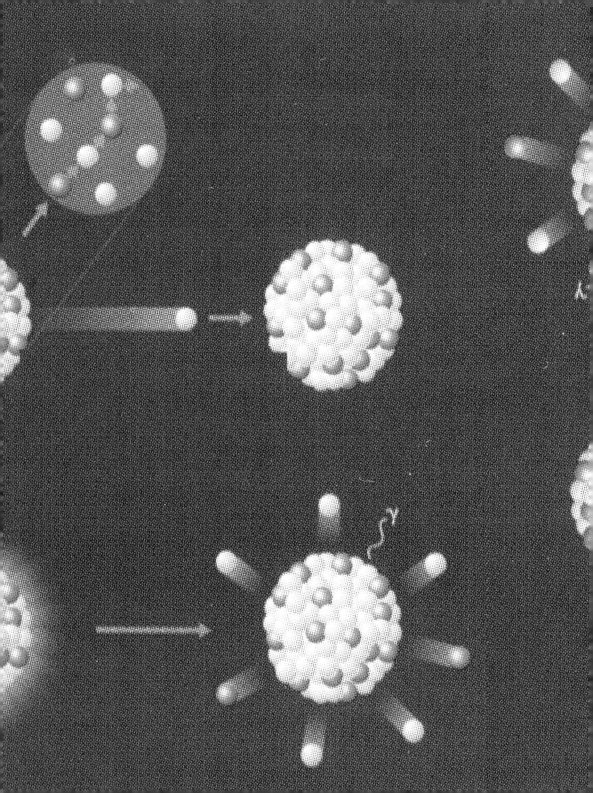

13

Complementarities
in the Øresund Region

The Øresund region is a divided region. It is not only divided by water but by the boundary between two sovereign states. These *cross-border regions* which are to be found in various parts of Europe are often considered to be peripheral from a national perspective. As a result of the centralisation of governing and managing capacities, in public as well as private sectors in the post war world, these border areas have been seen as relatively weak and backward from a central national point of view.

Changes are now on the way. There is considerable evidence to show that the propensity to change in the new Europe of the twenty-first century is greatest in these peripheral regions. In a situation where the central authorities of sovereign states are in the process of losing some of their former hegemony at the same time as the European Union is expanding, it would seem probable that the consequences of European political and economic renewal will be experienced most intensively in these border regions.

Looking for opportunities

The aim of this book has been to present and clarify the *opportunities* for development provided by a greater cross-border collaboration in the Øresund region. We have not dwelled on the *barriers* to integration that still exist since they have been widely discussed in other studies as well as in the mass media. It is important to emphasise here that the type of development on which we have concentrated here has related to qualitative growth, i e economic transformation and *renewal* rather than quantitative growth in the form of increasing population and urban concentration.

The aim of this book has been to present and clarify the opportunities for development provided by a greater cross-border collaboration in the Øresund region.

The choice of a European Spallation Source (ESS) as the starting point for our analysis and discussion has been made on several different strategic grounds. A scientific centre of this size and nature will almost certainly initiate processes of change in its local and regional environment. It is therefore an ideal object of study for an analysis of the socio-economic impacts of a regional investment project. As a research centre it will also place

The choice of a European Spallation Source (ESS) as the starting point for our analysis and discussion has been made on several different strategic grounds.

specific demands on its environment. It is on the basis of these premises that we have examined the resources concentrated in the neighbourhood of Øresund.

But the choice of a major regional investment project as the focus for our discussions should not hide the fact that local and regional transformation will in the long run probably benefit from a large number of individual initiatives and smaller projects and their effects taken together. It is also important to note that a large part of the theoretical and empirical analysis presented here may be applied to a wide range of investments in research and development projects irrespective of the actual scale.

A world in transformation

Two elements of the present transformation of industrialised economies have been of decisive importance when assessing the future prospects of the Øresund region. Firstly, account has to be taken of the growth of the *knowledge-based economy*. Secondly, the *regional level* has acquired greater economic and political significance.

Whereas previous periods of modern history have been characterised by a relatively slow transition from agrarian to industrial forms of production and living conditions, the present transformation is unique in human history in terms of the pace at which technology, forms of production and economic, social political and cultural conditions are changing. At the same time the fundamental importance of the *growth of knowledge and diffusion of innovations* has become increasingly evident.

There are several signs of the emergence of a post-industrial society and knowledge-based economy. *Trade* between industrialised countries has changed character. A growing proportion of the internationally traded goods comprises products whose high value has been created by inputs of highly skilled labour and advanced technology. The service content of these traded goods has increased while the pure material content has undergone a corresponding decline. *The content of work* and the production output have taken on new forms. Research has become the most expansive sector.

One of the most prominent features in the current development of advanced societies is the increase in the *number of years spent in formal education.* Sweden's transformation from one of Europe's poorest countries to one of the continent's richest has been accompanied by an increase in the average number of years spent in formal education from three to eleven years per employee. The major expansion of higher education belongs to the post-war world. In Sweden, for example, as late as the 1930s and 1940s, less than one per cent of an age cohort had received higher education. By the late 1990s, the corresponding figure was 40 per cent. Just after the Second World War there were less than 40 000 university graduates in Sweden. Today this figure is around one million. Postgraduate education has also expanded but not to the same extent.

The important role played by research, especially the *practical importance* to a nation of research in the natural sciences has been widely recognised. The *status and prestige of scientists* has risen markedly and has come to encompass not only natural science and medicine but research in general. At present there would appear to be a widespread impression that universities are an important driving force underlying technological and industrial development and that there is a fundamental relationship between research and higher education on the one hand and the international competitiveness of firms on the other.

This relationship would naturally have important consequences for employment and wealth of nations.

> It has long been known that there are agglomerative forces that exert a major influence on the location of different types of production. Firms and institutions gain advantages from proximity, particularly in certain areas and in certain places. It is especially important to note in the light of the aims of this report that the tendencies towards agglomeration in certain regions are even more striking in the knowledge-based economy than in the traditional industrial society with its factory towns and mining communities. *Geographical clustering* is one of the most prominent features in the economic geography of the new millennium.

Regions are also growing in importance in political terms. Three different types of regionalisation may be distinguished. One form is associated with *decentralisation* where power and competence that have previously belonged to the national level now are shifting to the regional and local levels. This type of decentralisation is to be found throughout Europe. EU's subsidiarity principle may be seen as a form of supranational support for this change. A second form of regionalisation is closely associated with cultural conditions and deeply rooted historical identities. This type of regionalisation is clearly based on an opposition to a national central authority; in many areas it has justifiably become known as *separatism*. The third and final form of decentralisation may be termed *region building* and is best suited to a situation where local and regional forces consciously strive to create a region that does not yet exist or perhaps more accurately try to strengthen an already existing but underdeveloped regionalism in an area. It is this latter category that best fits the case of the Øresund region.

A picture emerges of a fragmented space that does not feel entirely familiar, a Europe consisting of an archipelago of self-aware regions bound together by different types of network.

A picture emerges of a fragmented space that does not feel entirely familiar, a Europe consisting of an archipelago of self-aware regions bound together by different types of network. Several conditions have interacted to create this map. The operation of modern transport systems facilitates the development of nodal settlement systems. Strong forces of agglomeration encourage the concentration of productive activities. As we have seen above, firms and institutions within research and cultural activities are embedded in regional environments where human beings live and work.

However the lack of opportunities to exchange ideas, knowledge and capital over long distances would impede the growth of entrepreneurship, research and cultural diversity and threaten the individual region with stagnation. Extensive networks become established without an obvious consideration for national boundaries. The interaction between global forces of change and regional development ambitions is facilitated by the links that bring together a kaleidoscopic world of home bases and places of creativity. *Regions and localities do not disappear but become integrated in international networks that link up their most dynamic sectors.*

Universities in particular act as strategic links between worldwide networks and local environments. These links communicate in two directions. The university links up a place and a region with centres of knowledge throughout the world. They act as international connection centres. At the same time, the university mobilises local and regional competence in different ways to create an attractive environment in those places where they are situated.

Renewal attributes

Under the heading of "Cultures of Creativity" we discussed in a previous chapter, some of the features that characterise environments and milieux where epoch-making events had taken place and inventions had been made within for example art, architecture, music, philosophy, science and technology. As an example of such environments, we suggested areas, places and specifically local milieux as well as contact networks, organised meetings and institutions within which markedly creative people had worked. We found characteristics that appeared again and again. Before going on to examine the special features of the Øresund region in the light of these recurrent characteristics, let us first briefly summarise the arguments.

In all creative environments, there are human beings that have more or less unique competence. It is those individuals who are part of a physical and institutional milieu or belong to a network that basically create the preconditions for renewal.

Hence it is not actually the environment per se that is creative but the persons who are for various reasons to be found there. It is for this reason that we use terms like cultures of creativity and places of creation.

With a few exceptions, competence is based on a long tradition of knowledge and extensive experience. The pioneers are usually well aware of the achievements of their predecessors and look on themselves as the latest link in the long chain of knowledge. Different places seem to be associated with different traditions. A closer examination of these traditions reveals that these competent individuals have been imported into these areas rather than having their roots there. Milieux of creativity ought therefore to be seen as places and groups that attract human beings possessing unique competence within different areas. The tradition is partly a question of the same places and institutions being attractive during a long time. It has also been demonstrated that the periods of history characterised by radical transformation and renewal have also been periods of great mobility for artists, researchers and other intellectuals.

Diversity and variation would appear to favour creative processes whereas standardisation, uniformity and homogeneity do not. Many of the examples from the literature give an impression that a creative environment is almost chaotic. It is important to note that creative processes and radical renewal are often generated once unique competence and close communication coincide with instability and uncertainty. There is much evidence to support the view that all creative processes – applicable to technological inventions, research at the cutting edge or new art – involves the more or less systematic use of what may be termed "structural instability".

Even in well planned science parks and development centres, it has proved difficult to go beyond the preparatory stages of the actual creative process. The components that one believes to be necessary are gathered together in a small area. What happens next is difficult to determine. To an outsider, it appears to be largely random. Occasionally something interesting happens in this contrived environment. However, often the synergy effects are largely absent i e the results are no more than the sum of the parts. Each separate operation could presumably have generated these results without the major investments in the centralised facilities.

The studies that have been carried out confirm that it is very difficult to deliberately try to construct a creative milieu. On the other hand, it is fairly simple to destroy this type of environment by means of regulations and controls. Looking in the rear mirror, it seems that many of the innovations have been introduced *against* the intentions of the established institutions and organisations. There would appear to be a marked resistance to the introduction of new ideas in large organisations. Certain studies indicate that successful programmes of change have been implemented once the existing formal organisational structures have been overcome or indeed in some cases actually deceived by the innovators. Here the innovators have gone far outside the formal limits of the organisations in order to acquire information, resources and support. They have constructed coalitions and networks outside the formal hierarchies.

Viewed from this perspective, it is evident that communication between individuals and areas of competence is of strategic importance in a creative process. In the initial stages of a creative process, direct personal contacts and unexpected meetings that lead to new combinations of pieces of information would appear to be vital ingredients. In both geographical and institutional environments as well as in networks, *meeting places* are required for more or less random contacts. In relation to renewal, contacts that lead to new combinations of pieces of information are naturally neither predictable nor able to be planned in advance. Communication using advanced technology will come into its own once a contact net is established.

Blocking the exchange of information and closing the information channels is a highly effective way of destroying a creative environment. The meeting places that prove to be of the greatest importance are surprisingly often to be found outside the formal institutions and organisations where the professional and commercial competition is unable to block the free exchange of information.

On numerous occasions in this report, we have emphasised that the world of science is basically a powerful system of communications. Research and the distribution of knowledge assume that ideas spread and information circulates. Pieces of information combine in a new, frequently unexpected fashion in a creative process. By means of teaching, these new ideas will be passed on to others. The scientific community with its networks is not only used as a means of communicating ideas. It also serves as an instrument for control, criticism and recognition. At conferences, working meetings and symposia, individual networks are connected.

Bibliometric studies also show how written material disseminates information and facilitates control. The collections of letters written by authors, Nobel prize winners and other cultural figures are an excellent source for analysing communications networks right up to the time when e-mail became the primary source of correspondence. The letters written by Nils Bohr that are registered at the Institute that bears his name in Copenhagen provide us with a good example, demonstrating clearly his central role in the physicist network during the inter-war years.

Research and the development of competence

At the beginning of the year 2000, there were more than 130 000 enrolled students together with a further 10 000 researchers at the departments of the universities in the Øresund region. In a European perspective, this is a fairly modest figure. However, this concentration of research and higher education is the highest in the Nordic countries. By comparison, Stockholm and Uppsala have only half as many students and researchers.

In this report, we have chosen to use the number of publications in approved academic journals and conference proceedings as a measure of the volume of research in particular disciplines. From the mid-1990s, there are measures available for the number of published research reports in relation to GDP for a range of countries. Here Sweden is in second place after Israel and on a level that is twice that of the USA and far in excess of the average for the industrialised countries. The academic journals studied here belong to the fields of science, engineering and medicine.

However these nationally based comparisons miss an important point. *There are actually only a very small number of regions in different countries that provide high quality research.* In order to find empirical support for this argument, we referred to a Danish report in which bibliometric statistics had been collated on a regional basis, covering 39 urban regions in Europe all of which have nationally important universities and research departments. The source for the Danish study is a data base containing data from 5 000 leading academic journals in science, engineering and medicine.

There are actually only a very small number of regions in different countries that provide high quality research.

According to this study from the mid 1990s, measured in *absolute figures,* London, Paris, Moscow and the Amsterdam-Haag-Rotterdam-

Utrecht conurbation are the most important research regions in Europe. These are followed by Copenhagen–Malmö–Lund and Stockholm–Uppsala which is fairly remarkable viewed against the background of the geographic distribution of students enrolled at European universities. In *relative terms*, taking into account that urban regions vary markedly in size, Cambridge emerges as the most successful research centre in Europe with about double the amount of academic production per capita compared to its neighbour Oxford–Reading. The position of Geneva–Lausanne is tied up with activities of the CERN research laboratory. Stockholm–Uppsala, Helsinki and Copenhagen–Malmö–Lund are highly placed while large densely populated regions such as London, Paris and Moscow are far down the list.

Research in the fields of clinical medicine, bio-medicine, biology, biochemistry, biotechnology, ecology and environmental technology are often presented as areas of special competence in the Øresund area. Approximately 60 per cent of the entire pharmaceutical industry in the Nordic countries is located in the Øresund region. IT companies of varying sizes are also well represented in the Øresund area. It is also important to note that large scale and successful applied research has at the same time been carried out within the business sector.

The gainfully employed population on the Danish side of Øresund is more than twice the corresponding population on the Swedish side. This is particularly marked in relation to typical urban services such as bank and insurance, public administration and public services, transport, trade, hotels and restaurants. The difference between the manufacturing sectors on both sides of the sound is not as marked as in relation to services. Employment in agriculture and forestry is more or less the same on both sides.

During the first half of the twentieth century, manufacturing employment expanded in traditional sectors on both sides of Øresund such as textiles, clothing, shipbuilding and food-processing industries related to agriculture. During the 1950s and 60s, these sectors went into severe decline to be replaced by the more urban-based services. In addition, the composition of manufacturing industry has undergone changes.

It has been particularly important for this report to examine the industrial sectors that are most heavily involved in research and development. New firms and the research-intensive manufacturing industries recorded remarkable growth rates in the Stockholm region during the 1990s. This rapid rate of expansion was not encountered to anywhere near the same extent in the Øresund region. However this pattern is beginning to change.

While Stockholm would appear to be experiencing saturation, the research-intensive industrial sector is expanding, albeit from a low level, on the Swedish side of Øresund, particularly in Lund. On the other hand, this sector is not undergoing the same growth in the Copenhagen area. Here IT services have expanded markedly. These advanced knowledge-based services are especially demanded by both the university research sector and those industrial sectors that have become leading innovators.

Diversity and variation

There are not only disadvantages with a divided region. Among the advantages, there is the possibility that the two parts of the region may be able to complement each other in a quite different fashion compared to the situation that currently prevails. Let us examine a number of examples of what is termed *complementarities* in the geographical interaction theory.

Copenhagen may be characterised as a major city. By Scandinavian standards, it is a large service centre providing a wide range of cultural and recreational activities. On the Swedish side of Øresund, there is more available space, pastoral landscapes and even nature. The housing markets in different parts of the Øresund area complement each other. Since the Øresund and the national boundary raise numerous barriers, economic and social development have to some extent moved in different directions. This is not just a matter of national identity, language and legislation. Researchers have been able to show that there are considerable variations regarding culture, life styles and institutional conditions within public administration, education and the economy. In

Researchers have been able to show that there are considerable variations regarding culture, life styles and institutional conditions within public administration, education and the economy.

addition substantial immigration has taken place, not least on the Swedish side of Øresund. In a Nordic perspective, the cultural diversity in the Sound region is large.

The differences that emerge as a result of this diversity should not only be seen as giving rise to problems. The current debate on these issues would sometimes seem to suggest that this is the case. Progress cannot be expected in a world of uniformity and homogeneity. "Little progress would be made in a world of clones." It is quite within the realms of possibility that several of the 4 000-5 000 scientists who visit an ESS centre each year might consider cultural diversity and variation to be a substantial asset. They can themselves be expected to represent a wide range of cultural identities and life styles.

Communications

The geographical concentration of economic activity is a feature of the expanding knowledge-based economy. Our principal argument which is well supported by research findings presented in this book has been that the development of location patterns in this fast emerging economy is influenced by the need for personal contacts, particularly face-to-face contacts among highly educated specialists in research and knowledge-intensive business sectors. This may seem somewhat paradoxical given the communications technologies that characterise our age. However as we have seen above, this is related to special conditions affecting social communication in the renewal and innovation processes.

Good internal and external communications are one of the most important conditions to enable a region to become an important research centre. Efficient passenger transport systems are required in order that a region may be accessible to other development centres around the world. Travel opportunities within the region are also of major importance.

The estimates presented in the report indicated that the good transport facilities in the Øresund region make it one of the most central city regions in Scandinavia, viewed from a European and perhaps even global perspective. Among the 97 city regions that were compared with each other, the ones deemed to be more central were to be found in a banana-shaped arch from Great Britain in the north-west to northern Italy in the south. Given the fixed link between Copenhagen and Malmö together with the associated expansion of the road network in the region, internal travel communications may be said to be satisfactory.

All in all, the physical possibilities for contact would appear to be very good in the Øresund area. The extent to which this infrastructure will be used depends on how our need for contact develops in the future. In this context, we have emphasised the importance of institutional conditions and social networks. The concept of the social web or fabric has been used to denote the closely woven network contacts that bind together individuals in the world of research with key persons in the world of business, politics and public administration. This type of fabric has been shown to be the ideal basis for regional renewal and development. This is borne out by studies of successful regions and places around the world. Early on in this report, we pointed to several of the barriers to the formation of cross-border networks in the Øresund region. An ESS centre or other strategically placed research-intensive activity would however help to accelerate the process of integration, a development desired by many political commentators and interested parties.

The estimates presented in the report indicated that the good transport facilities in the Øresund region make it one of the most central city regions in Scandinavia, viewed from a European and perhaps even global perspective.

Strategic positions in the Øresund region

A detailed analysis of the various alternative locations for a major research centre within the region has considered to be outside the framework of this report. A number of jigsaw pieces that are important for prospective analysis of the ESS project may however be mentioned here.

As was emphasised at the outset, our aim has been to present various ideas and approaches that would enable us to have a broad discussion of the probable economic and social impacts of the location of research-intensive establishments. At the same time, the question has been raised regarding the demands that this type of large project may place on local and regional resources. The choice of an ESS centre as an example and point of departure for this discussion should not conceal this ambition.

The Øresund region has features that make it a relatively complete city region. In addition it is a major research metropolis in the European city landscape. The region has good communications facilities. Jointly these should make the region attractive to advanced scientific activities and to industries dependent on research and higher education. Finally returning to the ESS centre as an example, it is obvious that it has specific characteristics that place special demands on its environment.

The neutron source comprises a one kilometer long accelerator runway. This would require a land area of at least 1.3 square kilometres containing certain special features. The ESS is not a reactor. This is not a question of nuclear fission as at Barsebäck. High energy electro-magnetic radiation will not be able to escape from the plant. Nor is there any risk that the process will run amok. No warnings have been issued regarding the risk of a melt-down. On the other hand, the size and nature of the plant suggests that it can hardly be located close to centres of population.

If we nevertheless assume that it is the Øresund region that is the primary area of interest, the choice of location becomes limited. There is obviously more suitable land available on the Swedish than on the Danish side of the Øresund. Moreover the choice of a Swedish location would contribute to a better regional balance in the region. Currently and for the forseeable future, the centre of the Øresund region will be on the Danish side of the Øresund. This is attributable to the physical and cultural infrastructure of Copenhagen together with those features that accompany its roles as national centre and hub of international contact networks.

It remains to look for suitable sites in south-western Scania. Prior to a final discussion of a suitable location within this restricted area, let us bear in mind three specific locational advantages possessed by the Øresund region. Firstly there is the diversity and variation of a *major city*. Secondly a *major airport* provides international accessibility to an area on the northern periphery of Europe. Thirdly there is the *research and academic environment* concentrated in Lund on the Swedish side.

Within the framework of the geographic restrictions, there would appear to be two locations *to be weighed against each other.* In the long run, it is likely that the international airport at Kastrup on the Danish side will have to be complemented by an expansion of traffic at Sturup on the Swedish side (see Figure 2 p 16). In order to utilise both airports in a parallel fashion, it may be necessary to build a rail link for rapid transport between the two airports, using the bridge/tunnel link. Along this rail link, there will be a high degree of international accessibility. At the same time both Malmö, Lund and Copenhagen will be within convenient and easy reach.

The importance of proximity to the major research centre and university campus at Lund will have to be given considerable attention. A site between Lund and Malmö would offer proximity and excellent communications with Kastrup and Copenhagen. However this site must presumably be ruled out on the grounds of a lack of space. Considerations of space would suggest a site to the north-east of central Lund. This location would be close to the Lund Institute of Technology, the Ideon science park and a range of industrial establishments. On the other hand, this location is relatively peripheral to the central parts of the Øresund region and its airports. *These are some of the alternatives that need to be more closely examined in future studies.*

Appendix

ESS – THE SCIENTIFIC CASE[1]

The study of condensed matter within the disciplines of physics, chemistry, materials and life sciences provides the scientific basis for our society. Consequently, the ever increasing demands placed by society on material functionality in all fields of technological endeavour, from information technology, through civil, mechanical, electrical and electronic engineering to chemical engineering and medicine necessitate an ever deeper understanding of the fundamental structural and dynamical properties of materials at the atomic and molecular level. Such understanding is often elusive, and as a result condensed matter science and the complex phenomena that it embodies present some of the most significant intellectual challenges in both theoretical and experimental research.

Although condensed matter science by its very nature is "small scale" science, condensed matter scientists are increasingly meeting these challenges by turning to "large scale" neutron facilities. Indeed, the unique properties of the neutron (its deep penetration, its sensitivity to neighbouring elements, light elements and isotopic substitution, its magnetic moment and most importantly its unique kinematics that allow simultaneous determination of structural and dynamic properties of a material) make it an ideal tool for solving the problem of "where atoms are and what atoms do", as the 1994 Nobel Laureates Shull and Brockhouse so conclusively demonstrated in their pioneering neutron research of half a century ago.

The Role of Neutron Scattering

Neutron scattering is a sensitive and widely used experimental technique for studying the structure and dynamics of matter. However, it is a technique that is strictly limited by the intensity of available neutron sources. Indeed since the pioneering neutron scattering studies in the early 1950s this source intensity has increased only by a factor of four, an increase achieved almost 30 years ago with the commissioning of the world's premiere neutron research reactor, the Institut Laue Langevin in Grenoble.

Over the intervening decades the scientific and technological problems concerned with the detailed analysis of matter that depend crucially upon neutron scattering for their solution have grown in complexity, subtlety and range. As a response to this increasing demand for ever more sensitive neutron scattering methods *the European Spallation Source (ESS) Project* was established in the early 1990s as the most direct and feasible approach to providing a new third generation European neutron source. It is intended that the ESS will offer an effective increase in performance of between ten and a hundred over all existing neutron sources, depending upon the specific application. This achievement will represent by far the greatest single stepwise increase in source performance since the early 1950s.

The implications for neuron scattering science will be profound. Indeed, the role of the ESS in condensed matter science and technology will in many respects be analogous to that of the Hubble telescope in astronomy. The Hubble is changing our perception of the "outer universe", enabling us to see deeper and with greater clarity than ever before, elucidating phenomena that were previously at the limits of detection and revealing new phenomena beyond those limits. Similarly, the ESS will facilitate neutron scattering studies that will change our perception of the "inner universe", revealing new scientific phenomena and technological functionality through the deeper characterisation of the structural and dynamical properties of matter across all of the scientific disciplines Moreover, just as the Hubble telescope functions efficiently as part of a network of less powerful ground-based observatories, so the efficiency of the ESS will be enhanced by the supporting structure of the present less powerful European neutron facilities.

Flagship Areas

In order to visualize the so-called flagship areas, which may be tackled by the materialisation of ESS about seventy scientists from all fields of neutron science gathered during May 2001 in Engelberg in Switzerland. The goal was to deliberate upon the optimum choice for the neutron parameters of ESS and identify high profile research areas at the limit or beyond of what may be accessed today. The meeting resulted in a list of grand challenges within solid state physics, material science and engineering, biology and biotechnology, soft condensed matter science, chemistry, earth and related sciences, the science of liquids and glasses and the physics of the neutron itself. A conclusion was that ESS would lead to breakthroughs by three distinct methods:

- **allow scientists to address new problems, and to ask new questions**
- **provide new tools to tackle problems at the research frontiers**
- **offer high quality experimental data for unambiguous discrimination between theoretical models**

In the following, we briefly highlight an extract of the proposed grand challenges for ESS, which are representative of the topics listed above. Further details may be found in the Engelberg reports [1, 2]. The successful outcome of these challenges depends on the performance of ESS combined with the unique properties of neutrons as probe for the investigation of matter summarised in the section "Why Neutrons?" The section "What is Spallation?" describes shortly the spallation process of neutron production.

Solid State Physics

The interplay between neutron experiments and theory has driven the development of many new concepts in solid state physics and neutrons have played a pivotal role in the investigations of phase transitions and co-operative phenomena, magnetism and structure (static and dynamic) leading to advances in many other fields. The future challenges in fundamental solid state physics are the exploration and the understanding of collective behaviour of large numbers of interacting particles. Although future trends are notoriously difficult to predict, two important directions emerge.

Firstly the tendency to higher complexity, specifically materials which have physical properties determined by competing interactions, and secondly the trend to reduced dimensionality, both by synthesizing materials with low dimensional structural elements and by reducing the physical size of objects from the bulk to surfaces and interfaces, single atom wires and dots. Of basic interest in solid-state physics is to establish the ground state of relevant systems, which may be done by exploring possible excitations out of the ground state. For this, neutrons are versatile and often unique probes. Research at the ESS could lead to breakthroughs in quantum magnetism, many body physics, and other frontier fields.

In *magnetism*, significant advances are expected in synthesizing molecular and organic magnets, i.e. solids built from structurally well-defined clusters containing magnetic ions in a complex environment. These are of both fundamental importance and with respect to potential application in magnetic storage devices. New developments are also expected in exploring novel magnetic phases and their dynamics in low-dimensional systems. Exotic interactions, theoretically predicted, but hitherto unobserved interactions in solids will be discovered and exploited at ESS. There are numerous candidates for such interactions such as molecular magnets, high-T_c cuprates and f-electron compounds where higher-order interactions (e.g. quadrupolar, octupolar, three- and four-body exchange) are relevant, but their sizes could so far not be determined directly. In principle, neutron scattering allows the direct observation of higher-order term transitions, however, the associated transition matrix elements are typically two orders of magnitude smaller than for dipolar scattering.

The study of *phase transitions* will continue to be a major field of research with neutron based techniques. Systems of high complexity, exhibiting extreme many body effects (e.g. unconventional superconductivity) and low dimensional features are known and expected to undergo a large variety of phase transitions. Their exploration using neutron techniques will provide crucial insights into the microscopic mechanisms causing these phenomena. Of high current and most likely future interest is the relationship between spin polarization and transport of conduction electrons in specially tailored materials, *spintronics*. High intensity neutron beams will play a central role in elucidating the spin polarization and dynamics of these electrons. Experiments at the ESS will set new benchmarks for extreme conditions of external pressure and magnetic field, thus providing important insights into zero temperature phase transitions driven by quantum fluctuations.

Materials Science and Engineering

Neutron scattering has established itself as one of the most important tools for the analysis of materials. Metals, ceramics and their composites, semiconductors, superconductors, nano-phase materials, liquids, polymers, paints, lubricants, concrete, coal, wood, bones and biomaterials as well as many other materials have been analysed with great success using neutron scattering. As a result of these investigations the elastic properties of materials are very well understood, the danger from hydrogen embrittlement of steels can be predicted, and the dependence of the lifetime for metal – semiconductor junctions on the operation temperature can be determined.

The neutrons ability to penetrate many centimetres of engineering materials allows non-destructive studies of large components and samples in complex environmental or processing chambers. Neutrons are particularly well suited to non-destructive studies of components in their as fabricated and in-service condition such as engine parts and neutrons take a volume average such that results are relevant for the properties of real materials.

The sustainable growth of society can only be achieved if new materials and material combinations are explored on all length scales, time frames, and under real conditions. Many important areas have been identified as foreseen to play a vital role within the next decade and beyond and they will benefit tremendously from neutron scattering experiments at the ESS.

One example is in-vivo studies of the *functioning of lubricants* via the analysis of the macroscopic flow and the dynamics on a molecular level. Ineffective lubrication leads to permanent wear and failure of mechanical parts, which causes an estimated damage in the USA of about 6% of the gross national product. Neutrons can unravel the structure and dynamics of lubricants in moving engineering parts and understanding the dynamics on all length scales embracing macroscopic flow to molecular diffusion under real loads, will lead to the development of new lubricants for extreme conditions. Presently the lubricant layer in these measurements has a minimum thickness of about 0.3 millimetres, far more than realistic scales of industrial interest. Studies of reduced film thickness relevant to real applications (~10 micrometers) will become feasible at ESS.

Another example is *monitoring of switching times* in optical devices. Neutrons can monitor where the hydrogen atoms go and how they change the structure. With the ESS not only the steady state structure can be determined, but it will also become possible to monitor hydrogen penetration and diffusion in real scale fuel cells at operating temperatures. Yttrium metal coatings can be switched between reflecting and transparent by the charging and discharging of hydrogen. Newly discovered compounds promise faster switching times, as well as easy and safe hydrogen handling through electrolytic cycling. In the future, hydrogen will be crucial for energy storage materials and for energy conversion devices, which enhance the vital role of ESS for the understanding and optimisation of materials and of devices for future hydrogen based technologies.

The projected ESS intensity gain factors will have a huge impact on materials and engineering science and promote *time resolved experiments* with second to millisecond resolution; enhance spatial resolution to the important sub-millimetre region for residual stress, high pressure and high temperature experiments and for *monitoring of interfacial diffusion* and reactions at interfaces of metal and organic multilayers. ESS will also provide facilities for *in-situ real time experiments*, for example nucleation and recrystallisation of under cooled liquid alloys, the ageing and fatigue of alloys under cycling conditions; *routine determination of 3D-maps of stress and texture* within engineering components, neutron tomography for the production of 3D-images of machine parts under working conditions in real time, and with structural sensitivity and pulsed radiography to study fast time dependent phenomena with isotope sensitivity.

Biology and Biotechnology

Hydrogen and water are involved in all the molecular processes of life. Information on these aspects is seldom taken into account and this type of experimental information is mostly incomplete. In fact, until now the neutron research has largely been limited to some excellent scientific examples e.g. on the detailed *dissection of enzyme mechanism involving hydrogen* location. Because many enzyme reactions involve hydrogen there is great potential for wide application if the technical capability is improved. Moreover the role of water in *molecular recognition* is pivotal as, for example, the lubricant of protein ligand interactions or the bonding mediator. It is essential for neutron protein crystallography to find source, instrument and sample combinations to face this challenge.

Presently there are two major hurdles for wide application of neutron protein crystallography; firstly the size of crystals routinely available[2] and secondly a molecular weight ceiling; which for instance for the yeast genome have the result that at least half of all proteins in the genome are out of range of current neutron protein crystallography capabilities even if big crystals can be grown. With the projected gain at ESS the sample size requirement could be relaxed and numerous problems would become amenable for study by neutron protein crystallography methods.

Recently, a method has been developed to analyse wide-angle x-ray solution scattering data up to 5 Å resolutions in terms of the approximate positions of dummy amino acid residues. With neutron scattering, this approach can be extended to gain additional information about the *positions of individual residues*. This can be achieved either by the preparation of proteins with specifically deuterated residues or, for native samples, by making use of the change in contrast of residues during hydrogen/deuterium exchange, e.g. after placing a hydrogenated protein in heavy water. This requires a neutron source like ESS with high flux and high dynamic range and should contribute to the *determination of the protein fold in solution* from experiments on native samples. Furthermore, it would be an approach relevant for high throughput fold definition for proteins, which do not easily crystallise, e.g. detergent solubilised membrane proteins.

How will the opportunities at ESS map onto internationally agreed tasks and trends in the biosciences and open up new fields? In biology the post-genomic era, created by the speed and efficiency of gene sequencing, provides a huge stimulus to and the radical need for development of the structure determining techniques. It is a paradigm that structure is based on function. However, this paradigm has been recast largely by neutron inelastic scattering techniques to include dynamics. Thus structure and dynamics determines function. The major gains in capability of ESS for inelastic scattering should greatly widen the applicability to many more systems.

Biological membranes are worthy of a special mention. The extreme sensitivity of neutron reflection makes it uniquely suitable for the study of labile biological structures. The internal reflection at the solid/liquid interface combined with contrast variation allows the exact determination of the membrane structure and of the crucial polymer layer separating the biological membrane and the solid support. This provides *insight into the role of the soft polymer cushion* for maintaining membrane integrity and function, which is crucial knowledge for the design of advanced biosensors. Moreover, using 2D-detection and in-plane Bragg scattering, there is the chance to *study at molecular resolution the association and self-assembly of functional clusters* in the plane of the membrane. This knowledge is essential for the understanding of how proteins and lipids temporarily associate in a functional membrane.

A proposed flagship experiment is the *study of native membranes and whole cultured cell layers* on polymer cushioned solid substrates which permits the measurement of the cellular membrane response on the action of external stimuli (drugs, stress, pressure...) at molecular resolution and sheds light on the extremely poorly understood interaction between different membrane constituents under conditions of membrane transport, ligand receptor binding and cell adhesion.

Soft Condensed Matter

The concept of "soft matter" subsumes a large class of molecular materials, including e.g. polymers, thermotropic liquid crystals, micellar solutions, micro-emulsions colloidal suspensions and biological materials, e.g. membranes and vesicles. These substances have a wide range of applications such as structural and packaging materials, foams and adhesives, detergents and cosmetics, paints, food additives, lubricants and fuel additives, rubber in tires etc. and our daily life would be unimaginable without them.

Soft condensed matter systems in the future will increase in complexity both in structure and in the number and specific role of their components, e.g. multi-component soft and soft/hard materials tailor made for industrial applications. Such complexity will cover a wide range of length and time scales, posing challenging problems to basic science. Desirable systems show complex interaction potentials with several minima, generating different structures according to the mechanical and thermal history. The understanding of the *interplay of geometry and topology*, and the *characterisation of interfacial features* are of the utmost importance for the future developments and design of novel materials. Finally, the structural changes induced by external fields such as shear play a crucial role in the outcome of industrial processing. These issues have to be dealt with as a necessary precondition for achieving controlled improvement in the fabrication of future materials.

Neutron scattering in combination with advanced chemistry is the necessary tool for facing the new challenges in the field of soft matter. Here the focus is on linking chemical architecture to microscopic and macroscopic properties. The interplay between computer simulation and neutron scattering promises to become particularly effective because of the common ability of neutron scattering and computer simulation to home in on a key structural unit. Future trends will require a wide variety of experiments, including investigations on dilute components, or on very small amounts of matter such as particular topological points or at interfaces. In all these cases, very high intensities of the neutron beam are required. Assuming the availability of a high flux neutron source with the characteristics of ESS some flagship areas in this field can already be envisaged and include *rheology based molecular understanding* that will allow design of tailor made materials, the possibility of reflectivity studies of less transparent materials and thereby gain access to information about a whole new class buried interfaces.

Chemical Structure, Kinetics and Dynamics

With increasing neutron fluxes and particularly with the advent of the ESS, neutron diffraction experiments typically will involve the collection of a large series of diffraction profiles. These time-resolved, parametric experiments enable structural trends to be analysed as a function of physical parameters such as temperature, pressure and magnetic field. In turn, this leads to a fuller understanding of phase diagrams and structural transitions, and to deeper insights into structure property relationship. Time-resolved neutron diffraction studies are also very powerful means of following chemical reactions. Neutron powder diffraction allows bulk analysis of materials in "real-life" industrial configurations and yields important crystallographic, thermodynamic and kinetic information about reaction behaviour.

Environmental problems such as the green house effect, lead to research of new solutions for energy management. Fuel cells will probably be the cleanest and the most versatile power source of this century. However, many other scientific problems remain to be solved: efficient catalytic processes at electrode surfaces, in-situ studies of the evolution of catalytic reactions, identification of the active species and the reaction intermediates in catalytic processes, understanding ionic diffusion in solid-state electrolytes, chemical reaction kinetic optimisation.

Earth Science, Environmental Science and Cultural Heritage

The use of neutron scattering by the Earth sciences community has a relatively short history, but it is now obvious that *the potential of neutron scattering methods for the solution of Earth sciences problems,* including many environmental problems, *is enormous.* Many of the problems encountered in Earth sciences have, until recently, simply been too complicated for earlier neutron sources and instrumentation. Only with the advent of the latest generation powder diffractometers at modern spallation sources such as ISIS, has it become possible to study the crystal structures of minerals as a function of temperature and pressure with sufficient accuracy to be useful in solving subtle problems such as cation ordering. However, there are many areas in the Earth and environmental sciences for which the present sources and instrumentation are still inadequate. Examples include measurements of the *structural changes in minerals at* very *high pressures and* simultaneous high temperatures, locations of light elements in complex structures, *strain measurements and scanning of polycrystalline aggregates under non-ambient conditions*, and studies of the dynamical properties (neutron spectroscopy) also at non-ambient conditions.

This information will *enable the modelling of fundamental processes in the earth*, ranging from large-scale phenomena such as deep-focus earthquakes and volcanic activity, through to the transport (and disposal) of pollutants in the Earth's crust and stone preservation in monuments. With the foreseeable increment in neutron flux of a factor of thirty over ISIS, the ESS are expected to provide a much wider scope for these studies, hence opening a whole new area of Earth science studies. Further, the ESS will enable the tackling of many long-standing issues related to geological and environmental processes. The ability to construct sample environments that will reproduce the temperature and pressure conditions of the deep Earth, and the increase in neutron flux will allow in-situ studies of mineral behaviour and increase the understanding of the behaviour of the constituent materials of the Earth.

The contrast between the neutron scattering cross sections of atoms or cautions common in minerals, which have equal or similar numbers of electrons, makes neutron diffraction excellent for direct determination of site occupancies and order-disorder distributions not indirectly as when determined from x-ray data. Furthermore, although synchrotron x-ray resonant scattering can certainly be achieved to enhance scattering contrast in favourable cases, it cannot be performed systematically being dependent on available edges and bonding features.

To study the behaviour of minerals requires the reproduction of their "natural" environment and thus the need for simultaneous high temperatures and high pressures. In-situ studies are best suited for a thorough knowledge of the relations between thermo-baric variables and structural properties such as phase transitions, cation partitioning, bond valence, electronic structure, etc. Traditionally high pressures have been easier to work with using x-ray diffraction and diamond anvil cells, but the use of time-of-flight neutron techniques has lately allowed considerable progress in high-pressure mineralogy. The low attenuation of neutron beams by many materials can effectively make extreme sample environments (high temperature, high pressure, reaction cells, differential loading frames, etc.) easier to handle for neutron scattering than for other experimental techniques.

Phase and Texture analysis of natural materials such as stone and ceramics, is still relatively new but the potential application of such powerful techniques span many fields of interest within *archaeological research*, from standard fingerprinting to complex conservation problems. Owing to the non-destructive character of the techniques, their applicability to large, undisturbed objects and large volume of interaction as opposed to the surface analysis of x-rays, neutron scattering techniques are easily predicted to find many new applications in the fields of study and *conservation of historical artefacts.*

Fingerprinting often helps identify the actual quarries utilised in historical times and hence discover the original source of simple earth resources and thus investigate pre-historic trading and material exchange. In the non-diffractive mode, information on the inner fabric of large-scale materials and artefacts, which is beyond the reach of x-rays, can be obtained by making use of recently developed neutron detectors that lend themselves to neutron imaging and tomographic reconstruction. Applications of this technique to archaeological artefacts are already envisaged; the availability of improved instrumentation, especially in terms of detector capabilities, would definitely represent a major improvement for this new area of research.

Liquids and Glasses

Disordered materials play a central part in our daily life. Water covers two thirds of the Earth's surface and is the major component of our bodies. Glasses are in our windows, in optical fibres for communications and even eaten as candy or used as stable coatings on medicines. Ionic conductors are in batteries in cars (electric cars in the future), mobile telephones and computers. Yet our understanding of such materials, especially in relation to more ordered crystalline materials, is still very limited.

One of properties of neutrons that make them a key probe for the study of liquids and glasses is the ability to cover a large range of length and time scales. Other techniques, e.g. x-ray scattering, light scattering and nuclear magnetic resonance, can provide specific information over a wider range of either length or time, but not both. Even when the range overlaps the information obtained is nearly always additional to that from neutrons, not the same. As the trend is towards the study of more and more complex systems we envisage that neutrons at ESS will be used routinely as the central component in a study using multiple complementary techniques. Data will either be analysed simultaneously using sophisticated modelling techniques, or used as stringent tests of computer simulations. Such a coherent approach to the studies of liquids and glasses not only enables complete information from experiments, but also a detailed interpretation of the data.

A grand challenge for the ESS will be to *address one of the unsolved mysteries in the dynamics of amorphous solids and defect crystals*: two level systems or tunnelling states. These excitations seem to have a broad distribution at rather low energies. Their density is very low and thus inaccessible with present neutron sources, but they strongly influence the thermal properties, the heat conduction and other material properties.

Fundamental Physics

During the past twenty-five years, our world-view of nature has changed dramatically, ranging from the constituents of elementary particles to the status of the universe. Neutron physics has made major contributions to this evolutionary process of understanding. On the grand scale, cosmology has evolved into an exact science and neutron physics has contributed to the understanding of element formation and of phase transitions in the history of our universe. Various data extracted from measurements of neutron beta decay have been used extensively to fix the number of particle families at three. On the scale of the very small, neutron experiments have made substantial contributions to our under-standing of strong, electro-weak and gravitational interactions. Neu-tron interferometry and neutron spin-echo experiments have shown how non-classical states of neutrons can be created and used for highly sen-sitive investigations in condensed matter and fundamental physics re-search.

There are several examples of flagship experiments for ESS in Funda-mental Physics, one of them are sketched in the following. It deals with the question of *handedness of nature*. In nuclear decay experiments it was recognized in the late 1950s that one of the four fundamental forces - the weak force - is, as far as we have been able to discern so far, exclusively left-handed. Most Grand Unified Theories, however, start with a left-right symmetric universe, and explain the evident left-handedness of nature through a spontaneous symmetry breaking caused by a phase transition of the vacuum, a scenario, which, if true, would mean that the neutrinos today should carry a small right-handed com-ponent. Although limits on the right-handed currents have been de-rived from free neutron and muon decay experiments, what is really needed is a clear-cut "yes" or "no" experiment. Such an experiment, planned for ESS, is the two-body beta-decay of un-polarized neutrons into hydrogen atoms and antineutrinos, which occurs with a relative probability of $4.2 \cdot 10^{-6}$ compared to the usual beta-decay.

What is so interesting about this decay is that one of the four hydrogen hyperfine states cannot be populated at all if the neutrinos are com-pletely left-handed. A non-zero population of this sub-state would, there-fore, be a direct measure of a right-handed component. This experi-ment has severe background suppression requirements for which the pulsed structure of ESS, allied to its intensity, is well suited. Thus, with ESS it may be possible to prove for the first time that nature does not possess an intrinsic handedness and that there is exciting new physics beyond today's Standard Model of particle physics.

Why Neutrons?

Although complementary methods such as the other scattering probe, synchrotron x-rays, as well as local probes like nuclear magnetic resonance, electron paramagnetic resonance and Mössbauer, provide important information about matter, neutrons offer unique opportunities because of their very nature. The examples given in the previous sections all depend on one or more of the fundamental properties of the neutron of which the most important are:

- neutrons have the mass of hydrogen

- neutrons are neutral, element specific probes which interacts with the atomic nucleus in matter not with the electrons surrounding the nucleus, hence neutrons may distinguish neighbouring elements

- neutrons probe magnetism on the nanoscale because they have a magnetic moment that interacts with magnetic moments in matter with an interaction of the same order of magnitude as the interaction with nuclei

- neutrons interact relatively strongly with light elements and differentiates between isotopes in particular between hydrogen and deuterium and acts as an in vivo probe of hydrogen on the nanoscale

- thermal neutrons wavelengths comparable to characteristic length scales in matter and energies comparable to room temperature and below, hence thermal neutrons are well matched study to static's and dynamics of most terrestrial phenomena

- neutrons are highly penetrating, weak coupling with absolute and readily interpretable cross-sections

- neutrons penetrate deep into matter, are non-destructive and give no damage

What is Spallation?

Neutron sources have traditionally been based on reactor technology associated with the well-known problems of radioactive waste and high security risks. In contrast, ESS – the European Spallation Source – is based on the spallation effect, which produce neutrons effectively and almost eliminates the waste problem. Spallation is achieved when a fast particle, such as a high-energy proton bombards a heavy atomic nucleus; some neutrons are knocked out in the nuclear process called spallation. Other neutrons are boiled off as the bombarded nucleus heats up.

The process is a bit like throwing a ball into a bucket of balls with the result that a few balls are being ejected immediately while many more bounce around and fall out. For every proton striking the nucleus, twenty to 30 neutrons are expelled.

References:

[1] ESS-SAC/ENSA Workshop on "Scientific Trends in Condensed Matter Research and Instrumentation Opportunities at ESS", Engelberg/Switzerland, 3. to 5. May 2001. ESS/SAC/Report/1/01 edited by D. Richter.

[2] ESS-SAC/ENSA Workshop on "Performance of a Suite of Generic Instruments on ESS", Engelberg/Switzerland, 3. to 5. May 2001. ESS/SAC/Report/1/01 edited by F. Mezei and R. Eccleston. ISSN 1433-559X - ESS/INST/P1/01, May 2001.

Notes:

[1] Text written by Bente Lebech. Risø National Laboratory, Denmark.
[2] In today's structural biology research the sample size is rarely above 100 x 100 x 100 micrometers.

NOTES

1 Assignment

[1] The research programme is entitled *"The Role of Universities for National Competitiveness and Regional Development: The Case of Sweden in an International Perspective.* This project is headed by Sverker Sörlin, Swedish Institute for Studies in Education and Research (SISTER) and Gunnar Törnqvist, Department of Social and Economic Geography, Lund University. Kerstin Cederlund, Karl-Johan Lundqvist and Maria Wikhall from Lund University and Magnus Henreksson from the Stockholm School of Economics are among the participants in this project. The following international researchers have been involved:

Henry Etzkowitz, State University of New York

Raymond Florax, Free University, Amsterdam

Peter Hall, University College, London

Nathan Rosenberg, Stanford University

Sheldon Rothblatt, University of California, Berkely

A detailed presentation of the research programme along with some of the major research findings are to be found in Sverker Sörlin, Gunnar Törnqvist: *Kunskap för välstånd. Universiteten och omvandlingen av Sverige* . SNS Stockholm 2000. The final version of this report in English has the preliminary title Sverker Sörlin, Gunnar Törnqvist (Eds): *The Wealth of Knowledge.*

2 Background

[2] In the late 1970s, a comprehensive study was carried out into different forms of transport flows and contact networks between Zealand and Scania. These investigations were part of a series of studies that were designed to examine the question of a fixed link across Øresund. During the subsequent twenty year period, numerous statements have been made about changes in the underlying conditions. However remarkably few analytical studies have been carried out. The limited studies that have been made do not suggest that there has been a fundamental change that would allow us to talk about a growth of an integrated region during the 1980s and 1990s. Reference may be made to the following: *SOU 1978:20 Öresundsförbindelser. Konsekvenser för företag och hushåll. Bilaga B*; Gunnar Törnqvist, Folke Snickars: *Näringslivets Helsingborg – rastplats, marknadsplats, mötesplats*? Helsingborgs kommun, Helsingborg 1986; Åke E. Andersson, Christian Wichmann Matthissen: *Öresundsregionen*. Munksgaard/Rosinante, København 1993; G. Meyer: *Broen i vore hoveder. Identitet och vaekst i Öresundsområdet*. Handelshøjskolen i København 1997. The latter report along with appendices provides a relatively up to date account of how representatives of the business community view the need for contacts in the Øresund region.

[3] Peter Maskell: *Nyetableringer i industrien – og industrukturens udvikling.* Handels-højskolens Forlag, København 1992; ; Åke E. Andersson, Christian Wichmann Matthissen: *Øresundsregionen.* Munksgaard/Rosinante, København 1993.

3 ESS - European Spallation Source

[4] The appendix has been written by Bente Lebech at the Department of Material Research, Riso Research Centre, Roskilde. Anders Sjöland at Øforsk has edited the text.

4 Local and Regional Impacts

[5] A more detailed study of the earlier literature is available in for example Gunnar Törnqvist: *Arbetslivets geografi*, ERU rapport 3. Stockholm 1981.

[6] Gunnar Myrdal: *Rich Lands and Poor: the Road to World Prosperity*. Harper & Row, New York 1957; Gunnar Myrdal: *Economic Theory and Underdeveloped Regions*. Methuen and Co Ltd, London 1957.

[7] Francois Perroux: Note sur la notion de "pôle de croissance", *Economie Appliquee 8*, 1955; Francois Perroux: *L'Economie du XXeme Siecle*. Paris 1961.

[8] Joseph Schumpeter: *Business Cycles: A Theoretical Historical and Statistical Analysis of the Capitalist Process*. McGraw-Hill, New York 1939.

[9] Raymond Florax: *The University - A Regional Booster: Economic Impacts of Academic Knowledge Infrastructure.* Avebury, Aldershot 1992.

[10] This issue forms a central part of the research being carried out by Kerstin Cederlund at the Department of Economic and Social Geography at Lund University. This work is part of a research programme entitled "The regional roles of universities. Swedish education, research and regional development in an international perspective". It is financed by The Bank of Sweden Tercentenary Foundation.

[11] The following section owes its inspiration to an article by Charles Edquist "Systems of Innovation Approaches – Their Emergence and Characteristics" which is available in Charles Edquist (Ed.) *Systems of Innovation*: *Technologies, Institutions and Organisations*. Pinter, London and Washington DC 1997. Both the article and book contain a thorough discussion of relevant literature.

[12] Örjan Sjövall, Ivo Zander, Michael Porter: *Advantage Sweden*. Norstedts, Stockholm 1991.

[13] Charles Edquist & Bengt-Åke Lundvall, "Swedish Systems of Innovation", in R.R. Nelson (Ed), *National Systems of Innovation: A Comparitive Study.* Oxford University Press, Oxford 1993; Peter Maskell & Gunnar Törnqvist: *Building a Cross-Border Learning Re-gion*. Copenhagen Business School Press, Copenhagen 1999.

[14] Michael Porter : *The Comparitive Advantage of Nations*. The Macmillan Press, London 1990.

[15] Anders Malmberg, Örjan Sjövall, Ivo Sander: Spatial Clustering, Local Accumulation of Knowledge and Firm Competitiveness, *Geografiska Annaler* 78 B, No 2, 1996; Hans Tson Söderström (red.): *Kluster.se. Sverige i den nya ekonomiska geografin.* SNS Förlag, Stockholm 2001.

[16] Ron Martin, Peter Sunley: Deconstructing Clusters: Chaotic Concept or Policy Panacea? Submitted to *The Journal of Economic Geography*, December 2001.

5 The Knowledge-Based Economy

[17] The presentation below is based mainly on Sverker Sörlin, Gunnar Törnqvist: *Kunskap för Välstånd. Universiteten och omvandlingen av Sverige*. SNS Förlag, Stockholm 2000.

[18] Manuel Castells: *The Rise of the Network Society*, The Information Age: Economy, Society and Culture, Volume 1.Blackwell Publishers, Oxford 1996.

[19] Peter Hall, Pascal Preston: *The Carrier Wave: New Information Technology and the Geography of Innovation, 1846 – 2003.* Unwin Hyman, London 1988.

[20] Daniel Bell: *The Coming of Post-Industrial Society: A Venture in Social Forecasting.* Basic Books, New York 1973 and 1976.

[21] Gunnar Eliasson et al: *The Knowledge Based Information Economy.* Almqvist & Wiksell, Stockhom 1990.

[22] Manuel Castells: *The Informational City*. Blackwell, Oxford 1992.

[23] In addition to the works cited above see Manuel Castells: *The Rise of The Network Society*, The Information Age: Economy, Society and Culture, Volume 1. Blackwell Publishers, Oxford 1996; *The Power of Identity*, The Information Age: Economy, Society and Culture, Volume II Blackwell Publishers, Oxford 1997*; End of Millennium*, The Information Age: Economy, Society and Culture, Volume III. Blackwell Publishers, Oxford 1998.

[24] This data has been gathered from Sverker Sörlin, Gunnar Törnqvist: *Kunskap för välstånd. Universiteten och omvandlingen av Sverige.* SNS förlag, Stockholm 2000.

[25] Nathan Rosenberg, L.E. Birdzell: *How the West Grew Rich: The Economic Transformation of the Industrialised World.* Basic Books, New York 1986.

6 The Geopolitical Perspective

[26] Gunnar Törnqvist has worked on this material for the Nobel Foundation Jubilee Exhibition in 2001 together whith two doctoral students, Niclas Olofsson and Ola Thufvesson. See *Cultures of Creativity. The Centennial Exhibition of the Nobel Prize.* Science History Publications, USA & The Nobel Museum 2001.

[27] Statistical tables and maps have been carried out by Maria Wikhall at the Department of Economic and Social Geography in Lund.

[28] See especially Attila Varga: *University Research and Regional Innovation. A Spatial Econometric Analysis of Academic Technology Transfer.* Kluwer Academic Publishers, Boston 1998.

7 A Europe of Regions

[29] A more comprehensive definition and concrete examples are to be found for example in Gunnar Törnqvist: *Renässans för regioner. Om tekniken och den sociala kommunikationens villkor.* SNS Förlag, Stockholm 1998.

[30] Manuel Castells: *The Rise of The Network Society*, The Information Age: Economy, Society and Culture, Volume 1. Blackwell Publishers, Oxford 1996; *The Power of Identity*, The Information Age: Economy, Society and Culture, Volume II Blackwell Publishers, Oxford 1997; *End of Millennium*, The Information Age: Economy, Society and Culture, Volume III. Blackwell Publishers, Oxford 1998.

[31] The text in this section follows closely the arguments presented in *Europaperspektiv. Årsbok 2000*. (Eds. Ulf Bernitz, Sverker Gustavsson, Lars Oxelheim). Santerus Förlag, Stockholm 2000.

[32] Robert B. Putnam: *Making Democracy Work: Civic Traditions in Modern Italy*. Princeton University Press, Princeton N.J. 1993.

8 The New Economic Geography

[33] See for example Christopher Harvie : *The Rise of Regional Europe*. Routledge, London 1994; Jan deVet: Globalisation and Local and Regional Competitiveness, *STI Review*, 13, 1993, p. 89 –121; Philip Cooke: *Cooperative Advantage of Regions*. Unpublished paper, Centre of Advanced Studies, University of Wales 1994.

[34] Alfred Marshall: *Industry and Trade*. Macmillan, London 1919.

[35] Jane Jacobs: *Cities and the Wealth of Nations*. Penguin Books, Harmondsworth 1984.

[36] Paul Krugman: *Geography and Trade*. The MIT Press, Cambridge, Mass. 1991.

[37] See for example, Ash Amin, Nigel Thrift Eds: *Globalisation, Institutions and Regional Development in Europe*. Oxford University Press, Oxford 1994.

[38] Peter Maskell, Gunnar Törnqvist: *Building a Cross-Border Learning Region. Emergence of the North European Øresund Region*. Copenhagen Business School Press, Copenhagen 1999.

[39] Edward E. Leamer. Michael Storper: The Economic Geography of the Internet Age. NBER Working paper no. 8450. Accepted for publication in *Journal of International Business Studies*.

[40] Manuel Castells: *The Rise of The Network Society*, The Information Age: Economy, Society and Culture, Volume 1. Blackwell Publishers, Oxford 1996; *The Power of Identity*, The Information Age: Economy, Society and Culture, Volume II Blackwell Publishers, Oxford 1997; *End of Millennium*, The Information Age: Economy, Society and Culture, Volume III. Blackwell Publishers, Oxford 1998.

[41] Kerstin Cederlund: *Universitet – Platser där världar möts*. SNS Förlag. Stockholm 1999.

[42] Jean Labasse: *L'Europe des régions*. Gallimard, Paris 1991.

[43] Gunnar Törnqvist: *Sverige i nätverkens Europa. Gränsöverskridandets former och villkor*. Liber-Hermods, Malmö 1993 and 1996.

9 The Mechanisms of Regional Success

[44] *I framtidens kölvatten: Samhällskonflikter 25 år framåt*. Rapport från FA Rådet. Publica, Stockholm 1986.

[45] The research reports that are summarised here are presented in the following: Manuel Castells,

Peter Hall: *Technopoles of the World: The Making of 21ˢᵗ. Century Industrial Complexes*. Routledge, London 1994; E.Decoster, M.Taberies: *L'Innovation dans un Pole Scientifique et Technologie: Le Cas de la Cite Scientifique Ile de France Sud*. Universite Paris 1, Paris 1986; Peter Hall: *The University and the City*, *GeoJournal* 41.4, 1997; Peter Hall, M.Breheny, R, McQuaid, D.Hart: *Western Sunrise: The Genesis and Growth of Britain's Major High-Tech Corridor*. Allen and Unwin, London 1987; David Keeble: High -technology industry and regional development in Britain: The case of the Cambridge phenomenon, *Environment and Planning C* 1989; AnnLee Saxman: *Regional Advantage: Culture and Competition in Silicon Valley and Route 128*. Harvard University Press, Cambridge, Mass. 1994; Allan Scott: *Technopolis: High-Technology Industry and Regional Development in Southern California*. University of California Press, Berkeley 1993; S.M.Tatsuno: *The Technopolis Strategy: Japan, High Technology, and the Control of the Twenty-first Century*. Prentice-Hall Press, New York 1986.

[46] A summary is provided in Gunnar Törnqvist: *Renässans för regioner. Om tekniken och den sociala kommunikationens villkor*. SNS Förlag, Stockholm 1998.

[47] See for example Attila Varga: *University Research and Regional Innovation. A Spatial Econometric Analysis of Academic Technology Transfer*. Kluwer Academic Publishers, Boston 1998.

[48] Research in this area is being carried out at present by Kerstin Cederlund at the Department of Economic and Social Geography in Lund.

[49] AnnLee Saxenian: *Regional Advantage: Culture and Competition in Silicon Valley and Route 128*. Harvard University Press, Cambridge, Mass. 1994.

[50] This concept has been put forward by Torsten Hägestrand in different contexts including Torsten Hägerstrand: Resandet och den sociala väven, *Färdande och resande*. KBF, Stockholm 1995. The concept has been further developed in Gunnar Törnqvist : *Renässans för regioner. Om tekniken och den sociala kommunikationens villkor*. SNS Förlag, Stockholm 1998; Sverker Sörlin, Gunnar Törnqvist: *Kunskap för välstånd. Universiteten och omvandlingen av Sverige*. SNS Förlag, Stockholm 2000.

10 Cultures of Creativity

[51] The following text is a short summary of more extended works in Gunnar Törnqvist: LaCréativité: Une Perspective Géographique, *La Géographie de la Créativité et del'Innovation*. Université de Paris-Sorbonne, Paris 1989; Gunnar Törnqvist: Towards a Geography of Creativity in Ari Schachar, Sture Öberg (Eds) *The World Economy and the Spatial Organisation of Power*. Avebury, Aldershot 1990. See also *Cultures of Creativity. The Centennial Exhibition of the Nobel Prize*. Science History Publications, USA & The Nobel Museum 2001.

11 The Øresund Region in the European Urban Landscape

[52] We have previously referred to Edward E. Leamer, Michael Storper: The Economic Geography of the Internet Age. NBER Working Paper No 8450. Accepted for publication in the

Journal of International Business Studies.

[53] A more detailed account is provided in Gunnar Törnqvist: *Renässans för regioner. Om tekniken och den sociala kommunikationens villkor.* SNS Förlag, Stockholm 1998.

[54] Before continuing a discussion of the future of the Øresund region, it seems desirable to re-estimate all of these figures using current timetables. Such estimates are however time-consuming and will have to be postponed for the time being.

[55] Ulf Erlandsson & Chistian Lindell: Mittskandinaviens personkontaktmöjligheter med Europa, *Working Paper* No.3. Institutet för regionalforskning, Östersund. No year indicated.

12 Industry and Research in the Øresund Region

[56] A more detailed analysis is provided in *SOU 1978:20.* Öresundsförbindelser. Konsekvenser för företag och hushåll.

[57] The classification criteria were originally devised by Lennart Ohlsson, Lars Vinell: *Tillväxtens drivkrafter. En studie av industriers framtidsvillkor.* Industriförbubundets Förlag. Stockholm 1987. This classification has been used previously by Karl-Johan Lundqvist: *Företag, regioner och internationell konkurrenskraft – om regionala resursers betydelse.* Studentlitteratur, Lund 1996 and Gunnar Törnqvist: *Sverige i nätverkens Europa. Gränsöverskridandets former och villkor.* Liber-Hermods, Malmö 1993 and 1996.

[58] Karl-Johan Lundqvist, Lars Winter: Industrins struktur i Öresund. Status och utveckling, *Sjaelland och Skåne – före, under och efter bron.* Öresundsuniversitet 2001.

[59] Karl-Johan Lundqvist, Lars Winter: Industrins struktur i Öresund. Status och utveckling, *Sjaelland och Skåne – före, under och efter bron.* Öresundsuniversitet 2001.

[60] Sverker Sörlin, Gunnar Törnqvist: *Kunskap för välstånd. Universiteten och omvandlingen av Sverige.* SNS Förlag, Stockholm 2000.

[61] Karl-Johan Lundqvist, Lars Winter: Industrins struktur i Öresund. Status och utveckling, *Sjaelland och Skåne – före, under och efter bron.* Öresundsuniversitet 2001.

[62] Karl-Johan Lundqvist, Lars Winter: Industrins struktur i Öresund. Status och utveckling, *Sjaelland och Skåne – före, under och efter bron.* Öresundsuniversitet 2001.

[63] The data for the Öresund region is taken from *Öresundsuniversitetet: Uddannelse og forskning 2001.* Current information is available on the Internet at www.uni.oresund.org. It should be noted that the comparability of the data on researchers is uncertain.

[64] *Science and Engineering Indicators* - 1998.

[65] The material has been taken from Christian Wichmann Matthiesen and Annette Winkel Schwarz: Scientific Centres in Europee: An Analysis of Research Strength and Patterns of Specialisation Based on Bibliometric Indicators, *Urban Studies*, Vol 36, No 3, 1999.

[66] Håkan Westling *Iden om Ideon – en forskningsby blir till.* Lunds Universitets universitetshistoriska sällskap. Årsbok 2001.

[67] A more detailed up-to-date account of these and other coordinating bodies may be found on the Internet at www.oresundscienceregion.org

REFERENCES

Amin Ash, Thrift Nigel N. (Eds): *Globalisation Institutions, and Regional Development in Europe.* Oxford University Press, Oxford 1994.

Andersson Åke E., Wichmann Matthisen Christian: *Øresundsregionen.* Munksgaard/Rosinante, København 1993.

Bell Daniel: *The Coming of Post-Industrial Society: A Venture in Social Forecasting.* Basic Books, New York 1973 and 1976.

Castells Manuel: *The Informational City.* Blackwell, Oxford 1992.

Castells Manuel: *The Rise of the Network Society,* The Information Age: Economy, Society and Culture, Volume I. Blackwell Publishers, Oxford 1996.

Castells Manuel: *The Power of Identity,* The Information Age: Economy, Society and Culture, Volume II. Blackwell Publishers, Oxford 1997.

Castells Manuel: *End of Millennium,* The Information Age: Economy, Society and Culture, Volume III. Blackwell Publishers, Oxford 1998.

Castells Manuel, Hall Peter: *Technopoles of the World: The Making of 21st-Century Industrial Complexes.* Routledge, London 1994.

Cederlund Kerstin: *Universitet - Platser där världar möts.* SNS Förlag, Stockholm 1999.

Cooke Philip: *Co-operative Advantage of Regions.* Unpublished paper, Centre of Advanced Studies, University of Wales 1994.

Cultures of Creativity. The Centennial Exhibition of the Nobel Prize. Science History Publications, USA & The Nobel Museum 2001.

Decoster E., Taberies M.: *L'Innovation dans un Pôle Scientifique et Technologie: Le Cas de la Cité Scientifique Ile de France Sud.* Université Paris 1, Paris 1986.

Edquist Charles (Ed): *Systems of Innovation: Technologies, Institutions, and Organizations*. Pinter, London & Washington DC 1997.

Edquist Charles & Lundvall Bengt-Åke: Swedish Systems of Innovation, in R. R. Nelson (Ed): *National Systems of Innovation: A Comparative Study*. Oxford University Press, Oxford 1993.

Eliasson Gunnar et al: *The Knowledge Based Information Economy*. Almqvist & Wiksell, Stockholm 1990.

Erlandsson Ulf & Lindell Chistian: Mittskandinaviens personkontakt-möjligheter med Europa, *Working Paper* No.3.Institutet för regional-forskning, Östersund, no year indicated.

Florax Raymond: *The University - A Regional Booster: Economic Impacts of Academic Knowledge Infrastructure*. Avebury, Aldershot 1992.

Gibbons Michael et al: *The New Production of Knowledge. The Dynamics of Science and Research in Contemporary Societies*. SAGE Publications, London 1994.

Hall Peter: The University and the City, *GeoJournal* 41. 4, 1997.

Hall Peter, Breheny M., McQuaid R., Hart D.: *Western Sunrise: The Genesis and Growth of Britain's Major High-Tech Corridor*. Allen and Unwin, London 1987.

Hall Peter, Preston Pascal: *The Carrier Wave: New Information Technology and the Geography of Innovation, 1846-2003*. Unwin Hyman, London 1988.

Harvie Christopher: *The Rise of Regional Europe*. Routledge, London 1994.

Hägerstrand Torsten: Resandet och den sociala väven, *Färdande och resande*. KBF, Stockholm 1995.

I framtidens kölvatten: Samhällskonflikter 25 år framåt. Rapport från FA-rådet. Publica, Stockholm 1986.

Jacobs Jane: *Cities and the Wealth of Nations.* Penguin Books, Harmondsworth 1984.

Jönsson Christer, Tägil Sven, Törnqvist Gunnar: *Organizing European Space.* SAGE Publications, London, Thousands Oaks, New Delhi 2000.

Keeble David: High-Technology Industry and Regional Development in Britain: The Case of the Cambridge Phenomenon, *Environment and Planning C,* 1989.

Krugman Paul: *Geography and Trade.* The MIT Press, Cambridge, Mass. 1991.

Labasse Jean: *L'Europe des régions.* Gallimard, Paris 1991.

Leamer Edward E., Storper Michael: The Economic Geography of the Internet Age. NBER Working Paper No. 8450. Accepted for publication in *Journal of International Business Studies,* 2002.

Lundquist Karl-Johan: *Företag, regioner och internationell konkurrenskraft - om regionala resursers betydelse.* Studentlitteratur, Lund 1996.

Lundquist Karl-Johan, Winter Lars: Industriens struktur i Øresund. Status och utveckling, *Sjaelland och Skåne – före, under och efter bron.* Öresundsuniversitetet 2001.

Malmberg Anders, Sölvell Örjan, Zander Ivo: Spatial Clustering, Local Accumulation of Knowlege and Firm Competitiveness, *Geografiska Annaler* 78B, No. 2, 1996.

Marshall Alfred: *Industry and Trade.* Macmillan, London 1919.

Martin Ron, Sunley Peter: Deconstructing Clusters: Chaotic Concept or Policy Panacea? Submitted to *Journal of Economic Geography,* December 2001.

Maskell Peter: *Nyetableringer i industrien - og industristrukturens udvikling.* Handelshøjskolens Forlag, København 1992.

Maskell Peter & Törnqvist Gunnar: *Building a Cross-Border Learning Region. Emergence of the North European Øresund Region.* Copenhagen Business School Press, Copenhagen 1999.

Meyer G.: *Broen i vore hoveder. Identitet och vaekst i Øresundsområdet.* Handelshøjskolen i København 1997.

Myrdal Gunnar: *Rich Lands and Poor: the Road to World Prosperity.* Harper & Row, New York 1957.

Myrdal Gunnar: *Economic Theory and Underdeveloped Regions.* Methuen and Co Ltd, London 1957.

Ohlsson Lennart, Vinell Lars: *Tillväxtens drivkrafter. En studie av industriers framtidsvillkor.* Industriförbundets Förlag, Stockholm 1987.

Perroux François: Note sur la notion de 'pôle de croissance', *Économie Appliquée* 8, 1955.

Perroux François: *L'Économie du XXème Siècle.* Paris 1961.

Porter Michael: *The Competitive Advantage of Nations.* The Macmillan Press, London 1990.

Putnam Robert B.: *Making Democracy Work: Civic Traditions in Modern Italy.* Princeton University Press, Princeton NJ 1993.

Rosenberg Nathan, Birdzell L.E.: *How the West Grew Rich: The Economic Transformation of the Industrial World.* Basic Books, New York 1986.

Saxenian AnnLee: *Regional Advantage: Culture and Competition in Silicon Valley and Route 128.* Harvard University Press, Cambridge, Mass. 1994.

Schumpeter Joseph: *Business Cycles: A Theoretical Historical and Statistical Analysis of the Capitalist Process.* McGraw-Hill, New York 1939.

Science & Engineering Indicators – 1998.

Scott Allan: *Technopolis: High-Technology Industry and Regional Development in Southern California.* University of California Press, Berkeley 1993.

SOU: 1978:20. Öresundsförbindelser. Konsekvenser för företag och hushåll. Bilaga B till de danska och svenska öresundsdelegationernas betänkande.

Söderström Hans Tson (red): *Kluster.se. Sverige i den nya ekonomiska geografin.* SNS Förlag, Stockholm 2001.

Sölvell Örjan, Zander Ivo, Porter Michael: *Advantage Sweden.* Norstedts, Stockholm 1991.

Sörlin Sverker, Törnqvist Gunnar: *Kunskap för välstånd. Universiteten och omvandlingen av Sverige.* SNS Förlag, Stockholm 2000.

Sörlin Sverker, Törnqvist Gunnar (Eds): *The Wealth of Knowledge.* (Forthcoming)

Tatsuno S.M.: *The Technopolis Strategy: Japan, High Technology, and the Control of the Twenty-first Century.* Prentice-Hall Press, New York 1986.

Törnqvist Gunnar: *Arbetslivets geografi.* ERU-rapport 3, Stockholm 1981.

Törnqvist Gunnar: La Créativité: Une Perspective Géographique, *La Géographie de la Créativité et de l'Innovation.* Université de Paris-Sorbonne, Paris 1989.

Törnqvist Gunnar: Towards a Geography of Creativity in Shachar Ari, Öberg Sture (Eds):*The World Economy and the Spatial Oranization of Power.* Avebury, Aldershot 1990.

Törnqvist Gunnar: *Sverige i nätverkens Europa. Gränsöverskridandets former och villkor.* Liber-Hermods, Malmö 1993 and 1996.

Törnqvist Gunnar: *Renässans för regioner. Om tekniken och den sociala kommunikationens villkor.* SNS Förlag, Stockholm 1998.

Törnqvist Gunnar: Regioner och nätverk i framtidens Europa, *Europaperspektiv. Årsbok 2000.* (Red Bernitz Ulf, Gustavsson Sverker, Oxelheim Lars). Santérus Förlag, Stockholm 2000.

Törnqvist Gunnar, Snickars Folke: *Näringslivets Helsingborg - rastplats, marknadsplats, mötesplats?* Helsingborgs kommun, Helsingborg 1986.

Varga Attila: *University Research and Regional Innovation. A Spatial Econometric Analysis of Academic Technology Transfer.* Kluwer Academic Publishers, Boston 1998.

deVet Jan: Globalisation and Local & Regional Competitiveness, *STI Review,* 13, 1993.

Westling Håkan: *Idén om Ideon – en forskningsby blir till.* Lunds universitets universitetshistoriska sällskap. Årsbok 2001.

Wichmann Matthiessen Christian, Winkel Schwarz Annette: Scientific Centres in Europe: An Analysis of Research Strength and Patterns of Specialisation Based on Biblimetric Indicators, *Urban Studies,* Vol. 36, No. 3, 1999.

www.oresundscienceregion.org

www.uni.oresund.org.

Øresundsuniversitetet: *Uddannelse of Forskning 2001.*

IMAGES COURTESY OF:

Cover and Chapter 1, 2, 4, 5, 6, 7, 8, 9, 10, 11, 12
Formgivning Art2Di2 Image Archives

Chapter 3
ESS - Central Project Team
Geb. 09.1
c/o Forschungszentrum Jülich
D-52425 Jülich

Chapter 13
LUM No 6, June 2002 (http://www.lu.se/info/lum/Lum.html)